# Wildlife of Rogart

**Compiled by the Rogart Wildlife Group
following a two-year project**

200

*This project is part of the Highland BAP Implementation Programme, financed by the European Union under North and West Leader+ programmes (2000-2006), Scottish Natural Heritage and the Highland Council. We are also grateful for the support of the Sutherland Biodiversity Group, which is serviced by the Sutherland Partnership.*

# CONTENTS

**Page**

Introduction .................................................... 5

**Setting the Scene**
Parish, village and climate .............................. 7
Geology ........................................................ 9
Rogart: 18th and 19th centuries ...................... 11
Rogart in the 1940s: growing up on a croft ........... 18
Habitat and species survey, 2005 ...................... 20
Primary School: Wildlife Projects, 2005-2007 ........ 26

**Wildlife in Detail**
Vertebrates other than birds ............................ 29
Birds ........................................................... 38
Invertebrates and plant galls
    Introduction ............................................ 47
    Invertebrates other than insects ..................... 48
    Minor insect orders ................................... 52
    Bugs ..................................................... 56
    Beetles .................................................. 59
    Hymenopterans ....................................... 62
    Moths ................................................... 63
    Butterflies ............................................. 65
    Flies .................................................... 67
    Plant galls ............................................. 69
Microscopic freshwater life
    Animals ................................................ 73
    Desmids ................................................ 76
Flowering plants and ferns ............................. 79
Bryophytes .................................................. 101
Fungi ......................................................... 107
Lichens ...................................................... 116

Illustrations ........................... Between pages 72 and 73

# Introduction

The ***Sutherland Biodiversity Action Plan*** (2003) includes a section on 'Town and Village', but those involved in the preparation of the Plan found that there was little or no information available on the biodiversity of any of the towns or villages of Sutherland. An objective of this section of the Plan was, therefore, 'to raise awareness of the biodiversity on people's doorsteps', and a key target 'to encourage [up to five of] the larger towns and villages to undertake an audit of their wildlife'.

In February 2005, the Highland Biodiversity Project granted funds to the Sutherland Biodiversity Group for two Village Wildlife Audits, one at Scourie, on the west coast, and the other at Rogart, on the east. The report on Scourie was published in 2006; this book contains the findings for Rogart.

The Audit was launched at a public meeting in Rogart on 2nd March 2005 and, at this and subsequent meetings, the following formed the Rogart Wildlife Group: Andrew Coupar, John Macdonald, Morven Murray, Shirley Pearson, Malcolm Rider, Roger Smith and Fiona Winstanley, with Ian Evans representing the Sutherland Biodiversity Group.

The following public events were arranged during 2005: a bird walk in Corry Meadow on 22nd May, led by Alan Vittery; a wild flower walk along the river bank and canal at Davochbeg on 19th June, led by Ken Butler and Morven Murray; an insect walk in Corry Meadow on 24th July, led by Philip Entwistle; a bat walk from Pittentrail to Davochbeg on 3rd September, led by David Patterson; a fungus foray on Corry Meadow on 18th September, led by Ian Paterson. The attendance was good, averaging about ten.

These events generated much useful information about the wildlife of the area, but seemed only to have scratched the surface. We had asked Viv Halcrow to spend a day surveying habitats and species in the vicinity of the village, which she did on 11th August 2005. She concluded that **an exceptional range of habitats of value to wildlife occurs within walking distance of the village centre**.

During 2006, members of the Group continued to record aspects of the wildlife in which they had an interest, but we also asked a number of naturalists with particular expertise to have a look at the area. We were fortunate that members of the Highland Biological Recording Group, together with visitors from much further afield, were able to spend some time in Rogart during the year; the detailed sections of this book record their findings.

We originally intended to confine the Audit to six 1km squares centred on the crossroads at Pittentrail (NC 726020), but since Rogart is a rather 'sprawling' community, we later enlarged the area covered to adjacent parts of the 10km square NC70, and at times, even a little further afield.

We should like to thank all who have contributed to the Audit and the book that has resulted from it. In addition to those mentioned above and elsewhere in the book, we are grateful to Ken Crossan for the use of illustrations from his Sutherland Collection (held by the Sutherland Partnership), to others who have contributed illustrations, and to Ann Cook and Isobel Patience for their help in putting the book together.

The first part of this book is concerned with **setting the scene**, including contributions on historical aspects and work being done at the Primary School. The second part consists of accounts of the **wildlife in detail**.

We hope that you will enjoy reading about the findings of this Wildlife Audit, and that their publication will encourage us and others, in the future, to build upon them.

**Rogart Wildlife Group**

# SETTING THE SCENE

## Parish, village and climate  *Andrew Coupar*

The **parish** of Rogart lies in south-east Sutherland, in the north of Scotland. It extends to some 272 sq km (105 square miles). At its lowest point in the south-east, near Kinnauld (7400*), it is a mere 5m above sea level, rising to 502m at the summit of Meall an Fhuarain in the north-west (5923). The area is drained by two main river systems, the Fleet and the Brora, both of which flow to the south and east. The upper reaches of the Black Water, a major tributary of the Brora, mark much of the north-eastern boundary of the parish. There is a scatter of lochs and lochans in the parish, but none of these extends much beyond 50 ha. in area.

The land is predominantly moorland, with dry heath on the better-drained slopes, wet heath and bog on the more level ground and in the intervening hollows. Woodland, mostly dominated by birch, is largely confined to lower ground in the south-east, particularly, but not exclusively, along river corridors. Grassland is predominantly associated with areas of pasture, again mostly in the lower and more fertile south-east.

Although perhaps not at first apparent, all the land is used and managed, mostly for sporting purposes, particularly deer stalking and some grouse shooting, and as rough grazing. Farming and crofting on the lower, more fertile, ground is largely focussed on sheep, and, to a lesser extent, cattle, with a small amount of cropping.

The population of Rogart, at around 400 (compared with 2,022 in 1801), is also concentrated on the lower ground in the south-east, particularly in and around the **village** of Pittentrail (7220) on the A839, but extending over a number of smaller communities, such as those at Little Rogart (7204), Knockarthur (7506), Rhilochan (7407), Langwell (6908) in the north, and Tressady (6904), Muie (6704) and Acheilidh (6604) in the west. As explained in the Introduction, the Wildlife Audit concentrated on the area around Pittentrail, but did extend further afield at times.

Rogart's **climate** reflects its position in south-east Sutherland, together with its extent and altitudinal range. The summers, although sometimes a little late in coming, are typically warm, with reasonable sunshine, and in recent years at least, autumns have been long and mild. In winter, deep, drifting snow can affect the higher ground for lengthy periods. For its part, the lower ground, alongside the Rivers Fleet and Brora, suffers intermittent flooding at times of snow-melt and/or high rainfall.

Although there are no official weather stations near to Rogart, Jim Oliver has been keeping detailed records from his garden near the Village Hall for many years. For the year April 2005 – March 2006, the total rainfall was 905mm. The wettest month by far was November, with 140mm, the driest being July and February, with 27mm and 30mm respectively. As to the temperature, July was clearly the warmest month, with an average maximum of 19.6°C. January had the lowest average maximum, 6.1 °C, but, interestingly, March had the lowest average minimum, at –0.3 °C

[* **Grid references**. Since virtually all of Rogart lies within the 100 km grid square NC (29), we have frequently used the simplified form 70 or 7400 in this publication for the relevant 10km or 1km square. Ordnance Survey coverage of the area is provided by the 1:50,000 Landranger Series, sheet 16 (Lairg, Loch Shin and surrounding area) and the 1:25,000 Explorer Series, sheet 441 (Lairg, Bonar Bridge and Golspie).]

# Geology
*Malcolm Rider*

The extent of the relationship between the geology of an area and its present-day wildlife may be overlooked by the casual visitor, but it is often central to an explanation of why plants and, to a lesser extent, animals, occur where they do. At the largest scale, the geological history of a landscape determines its topography and the extent and nature of drainage patterns and waterbodies. These, taken with the mineral composition of local rocks, influence the character of the soils formed from these rocks, the vegetation the soils support and other plant, fungal and animal life associated with this vegetation. At the most intimate level, the mineral content of the rocks determines whether organisms such as lichens and bryophytes, which grow directly on their surface, can get a foothold and thrive.

The Moine rocks which underlie Rogart are part of an ancient metamorphic succession, which covers most of the Northern Highlands. However, the final sculpting of the hills and sedimentation of Strath Fleet took place during the most recent ice age, just a few tens of thousands of years ago, and by the sea even more recently.

Moine rocks are hard, crystalline and usually slabby. They were originally formed 1000 million years ago, but were heated, re-formed (metamorphosed) and twisted during the building of the Caledonian mountains, around 450 million years ago, as the ancient continents of Laurentia (to the north) and Baltica collided. The steep hillsides on the approach to the village from the A9 show these rocks nicely, as does the quarry at Morvich. They are mostly crystalline quartz, sometimes with dark, shiny, biotite mica layers, and often showing quartz veins.

Although the underlying rocks are old, the real character of Rogart was formed during the ice ages, principally from 25,000 years ago, when the area was covered by kilometre-thick ice glaciers, until about 8,000 years ago, when the melting ice was dumping its sediment in the Strath, and finally and much more recently, when the sea was lapping Rogart.

The ice gouged out Strath Fleet and ground down the hills to their present rounded outlines. This explains why the rivers that come from the sides of the Strath, like the Corry Burn, tumble down to the main river in steep, canyoned cascades. They are hanging valleys and formed to feed glaciers not straths.

After the ice came the floods. The sand, gravel and boulders that you see in the present River Fleet were washed-in from the melting ice. But the final phase has been influenced by the sea. After the ice melted, Strath Fleet was an estuary. Then, following this melt, as in the rest of Scotland, the land, relieved of the weight of the ice, began to rise. In Rogart, the flat areas 5m or so above the present level of the river and flanked by steep, gravelly slopes, are local evidence of where the sea level once was, perhaps 8000 years ago. But the flat, valley floor fields as at Morvich, also originally strand flats, are even more recent. Before The Mound was completed, in 1816, the sea came up almost to Rogart.

So although the rocks underlying Rogart are ancient, much of the scenery and the look of the fields today are historical.

[For more information visit: www.scottishgeology.com]

# Rogart in the 18th and 19th centuries   *John Macdonald*

The Rogart area has been occupied by man for over 5,000 years. Woodlands have been cut for timber, firewood and browse (and new ones planted), peat has been cut for fuel, land has been drained, watercourses straightened and embanked, some areas have been cultivated, most have been grazed by stock, rock has been quarried for building and road making. The River Fleet was tidal up to Pittentrail until the construction of The Mound in 1816. All of these activities, and others, have affected the landscape and the wildlife it supports.

We have little information on the wildlife of Rogart until the mid 19th century, but there are two accounts of the parish, from the late 18th and mid 19th centuries, which give us an idea of how the landscape looked then and how it was being managed. They are contained in the reports made by Ministers of the Church of Scotland for the *'Old' Statistical Account* organised by Sir John Sinclair during 1791-1799, and its successor, the *New Statistical Account* of 1834-1845.

The account of Rogart in the ***'Old' Statistical Account*** is by the Reverend Aeneas McLeod and dates from 1793. Relevant parts of it run as follows:

> *Name, Situation and Extent.* The name seems to be of Gaelic origin; and the parish to have been so denominated, from the peculiar situation of the parsonage, which is nearly on the top of a high hill, and thus *Rogh-ard*, i.e. *very high*, came to be written, as now, *Rogart*. It lies in the county of Sutherland, presbytery of Dornock, and synod of Sutherland and Caithness. Its extent is not above 10 miles, and it is nearly as broad as it is long.
>
> *Surface and Soil.* The surface is most irregular. The two straths, Strathfleet and Strathbrora, which make the principal part of it, are rather more regular than the rest; and yet even in these, very few acres of ground are found together, that are not intersected, either by brushwood, growing from old stocks of trees, or by very rapid destructive burns, running down the hills on both sides. These straths run from west to east the whole length of the parish, and lie 5 miles distant from each other. The interjacent space is a group of rocky hills, with moss intervening here and there. The sides of the hills are, for the most part, cultivated; but it is rare to see 3 yards of

ground without a rock. The soil is therefore thin, and almost everywhere mixed with blue sand, or gravel. The tops of the hills are commonly covered with a dry short heath of little value. In the strath, the soil, in general, is a light loam, and where the rivers do not overflow, not infertile.

*Climate, &c.* The climate is sharp and cold; the winds beat strong on many places, and, from the immediate neighbourhood of the lofty mountains of Strathnaver, much rain falls here, yet the people in general are healthy, no particular distemper being prevalent among them. They all have plenty of good peats, and lodge dry and warm in their houses. There are several persons living in the parish considerably above the age of 80. The ravages of smallpox is still to be lamented, inoculation having obtained but little among the lower class....

*Wood.* Some wretched vestiges of very considerable birch woods are to be seen in different parts, but the shoots from such of the old stocks as have not decayed, are annually cropt by the cattle in autumn and winter; and such shoots as may survive to a second summer, are sure to be cut by the people to feed their cattle.

*Population.* About 40 years ago [1753], when the returns were made to Dr Webster, the number of people in the parish was stated to be 1761. They now amount, by particular enumeration, to be about 2000....

Reverend McLeod continues with observations on *Occupations*: 'A great many people in the parish call themselves tradesmen, and, at times, exercise their several crafts, as weavers, taylors, shoemakers, smiths, carpenters, coopers &c. But there is hardly any one in all these professions that does not hold more or less land; so the whole may be said to be farmers, and their chief property consists in the number of cattle of different sorts they can keep.'

There follow sections headed *Heritors and Rent, Church, Poor* and *Crops and Expence of Labour*. We learn from the last that: 'The only crops in the parish, are oats, barley, and potatoes.' He then discusses labour problems, and goes on to castigate 'young boys, who are commonly very restless, running to the south of Scotland for higher wages during the warm season...and returning to live idle with their friends from November to March...their earnings scarcely compensate for the extravagance of dress, and other vices they bring home with them.'[!]

The report on Rogart in the *New Statistical Account* is by the Reverend John Mackenzie and dates from September 1834. In the intervening 41 years, the population had experienced a major upheaval. The old rig system had been abandoned and the whole parish re-arranged into individual holdings. The enclosure of these was still in progress. Medium sized farms had been created from the more fertile lands, and smaller farms from the more mixed pastures. A lot of people had been displaced to make way for the creation of extensive sheep runs and had either left the parish or been crowded in amongst the remaining population. They had had to accept bare hillside and set about taking this most unpromising land into cultivation. The report touches on the overcrowding of the population when it reports on schools: 'There is a district of the parish, *Barrschol* and *Craiggies*, containing a population of 200, which is four miles distant from the parochial school, and has no other school within reach.'

The following extracts from the 1834 report are perhaps the most relevant to this account of its present-day wildlife:

> *Topographical Appearances.* Strathfleet, in the language of the inhabitants, is called Strathfloid; and the small river passing through it is called in that language *Flodag* the diminutive of *Flod*, a word signifying inundation, to which this stream is subject. This strath is ten miles in length...Both sides of it rise to an elevation of from 500 to 700 feet above the course of the Fleet, in some parts abruptly, but generally in sloping banks, which are occasionally cultivated and produce crops.
>
> The part of Strathbrora that is within this parish bears a resemblance to Strathfleet, the difference being such as may be accounted for by the action of a larger volume of water, which has in some places cut deeper into the rock, forming chasms...Being nearer the mountainous region, the aspect of this strath is of a more rugged character than that of Strathfleet.
>
> The hills between these straths are nearly of equal height, and rise to an elevation of from 800 to 900 feet...The meadows, which are

found around some of the lakes and in those flat parts subject to irrigation from burns passing through, are not of great extent, and form but a small proportion to the extent of moors.

*Climate.* The climate...is sharp and cold. The greater part of Rogart...having but little shelter from the east wind, and being swept by every blast coming from the high mountains of Assynt and Strathnaver, is much exposed to the severity of a cold atmosphere. Yet snow does not lie long here, and frost is not very intense. Winter, however, leaves us but reluctantly, continuing during the greater part of spring; and it often arrives in the last month of harvest. The most frequent winds, however, are the north and east, but the south-west blows with greatest violence.

Summer here has a great proportion of dry weather; as the rains which fall among the high mountains in this season do not extend to this place. A dry scorching summer is more frequently a subject of complaint with us than one too rainy....Notwithstanding the coldness of our climate, however, it is remarkably healthy.

*Hydrography.* The lakes in this parish are very numerous, but not remarkable for extent. Of Loch Craggie, in its western extremity, anglers speak with rapture for the size and quality of its trout, and for the excellent sport it affords. In the north-eastern extremity of the parish, there are two lakes in which fine trout are found. Their name implies that they were once on this account held in estimation; both being called *Loch-Beannached, Lake of Blessing.*

The only rivers are those already mentioned, the Fleet and the Brora. Even the larger of these is insignificant in summer and harvest; but both when in flood, discharge a great body of water, and often cover the whole of the plains in their courses, so as to present the appearance of a sucession of lakes. The Fleet has its origin in a rising ground, forming the boundary between the parishes of Rogart and Lairg. After...many windings, fringed with birch and alder bushes, it enters an extensive plain, once covered by every tide from the Moray Frith, but now encroached upon only by this stream: the waters of the sea being now completely cut off by the earthen mound, at the head of the *Little Ferry.* In this place, where it is not confined by the skill and enterprise of the agriculturist, it appears almost completely lost among rapidly growing alders, until it collects itself into a pool, or forms a considerable lake, before being discharged into the sea by the sluices of the mound.

*Geology and Mineralogy.* Rogart lies chiefly on gneiss rock [not now so described], in which the only veins seen are of quartz. It is of a large-grained kind, with a great proportion of mica. It is used in building the houses and cottages of the inhabitants, and is found an excellent material for this purpose, being easily wrought. Over the whole of the parish, rolled blocks of granite are seen in great numbers on the surface; in some parts, if viewed from a distance, the surface appears covered with them. They are found no less numerous under ground in hollows, where there has been an accumulation of soil to cover them.

Of the whole surface of Rogart, moss [peat] forms the largest proportion. In some parts it is very deep, found often to a depth of twelve feet. In those parts where the depth is less, its fresh appearance indicates rapid growth. The soil in the valleys, and covering the sides of the hills is sandy and gravelly. The land abounds in springs; consequently, to be brought into a state of culture, it requires to be intersected with frequent drains.

*Plants.* The moors produce heather, deer's-hair and cotton-grass,. intermixed in proportions said to be highly favourable for the feeding of sheep. The hills are covered with heather on the tops, but on their sides a mixture of fine grasses is to be found; and around their bases, red and white clover, and mountain daisy, are common. The meadows and straths are covered with meadow grasses prevalent in similar situations, and, where irrigated, are very productive.

*Zoology.* Roe-deer may always be seen here, but not in great numbers. The red mountain-deer is occasionally seen crossing the moors to or from the mountains north of this, which abound in that species of animal. The gray mountain hare is here common on the higher grounds. The brown hare, and of late, the rabbit, are found on the lower grounds, the former exceedingly numerous. Moor-fowl [red grouse] are still abundant, though less so, it is said, than they have been. Black game, which are said to increase as moor-fowl decrease, are become very numerous.

Goats were once a part of the stock of the inhabitants, but they have now nearly disappeared, giving way to more profitable animals. There is a species of sheep, of small size, formerly the only kind known here, still reared by occupants of small lots of land, and much commended for fineness of fleece and excellence of mutton; but they are likely soon to disappear also, those who have them appreciating the better size of the Cheviot sheep.

A great variety of trout is found in the lakes. Salmon, grilse, and sea trout, are taken in the Brora and the Fleet. The trout make for the burns falling into, or issuing from the lakes, in the month of October, to deposit their spawn; and their spawning season lasts generally till the beginning of November, and seldom or never extends beyond the middle of that month. Salmon begin to spawn fourteen days later, and before the middle of December; that process being finished, they return to the sea. Salmon enter the Fleet in the end of May. They are found, and were taken, till a recent act of Parliament prohibited, at the mouth of the Brora, as early as the end of January; but they are not seen in the upper part of that river, the part belonging to this parish, till the commencement of summer.

*Agriculture and Sheep-Farming.* The proportion of the land in culture and yielding crops is small, and must always be so, while naked rock covers a considerable part of the surface of the parish. It has, however, since several years back, been increasing; and it is likely, that, in the course of some years, what is now occupied by lotters, if left in their occupation, may become cultivated where practical. The quantity of land cultivated, or occasionally in tillage, does not at present exceed 1200 acres.

Nearly half of the parish in value, and more than half in extent, is laid under sheep of the Cheviot breed. It is no less than 62,800 acres in extent

*Husbandry.* The pasture for sheep is good, and the sheep reared on it are said to be the best quality of their kind. Surface draining, which has been carried on to a great extent, has added much to the quantity, and improved the quality, of feeding for the sheep. In this species of improvement, little remains to be done here by the sheep farmer.

*Live-Stock.* The number of sheep of all kinds is 6420; of black cattle, (heads of,) 1079; of horses 276; of pigs 210.

*Woods.* Timber as yet cannot be mentioned as one of the products of Rogart. A small space in Strathfleet [Tressady], about twenty acres, having some native plants of oak, was enclosed and planted with larch and common fir. The appearance of this small plantation, which has been lately thinned for the first time, affords sufficient encouragement for planting in situations equally favourable... Small alders are to be seen along the streams; and patches of dwarf birch are common. Both of these, when in foliage, enliven the aspect, and relieve the ruder features of the scenery; but otherwise are of no value.

*Fuel.* Moss, cut as peats in the months of May and June, and abounding of the best quality at no great distance from the inhabitants, is the fuel used by all. It is procured at considerable expense of time and labour; but the very poorest never fail to supply themselves with a stock sufficient for the year's consumption.

The Reverend Mackenzie concludes with comments on the 'improvements which have taken place here, since the time of the last Statistical Account....Roads and bridges justly claim particular mention...They have changed the mode, as well as improved the facility, of every species of carriage. Sledges...were formerly the best means of carriage which those in better circumstances could use for farming, and for other purposes. Now, almost every poor man who cultivates a croft off land, has his wheeled cart.' He also remarks on the houses, 'In no part of the North Highlands, are there so many well built neat-looking cottages...and their number, seen as they often are, in picturesque situations, must strike every observer, as giving life and interest to the scene presented to his view. Whoever sees them, must form a favourable idea of the industry of the inhabitants, and of the encouragement afforded them by the proprietor of the soil.'

# Rogart in the 1940s: growing up on a croft

*John Macdonald*

Ours was the last generation to experience life on the croft prior to the coming of electricity, running water, television, PlayStations, iPods and what have you. But this did not deprive us in the least. Nature was all around and you were part of it.

Being young and inquisitive, we were always finding birds' nests. There were Skylark nests among the moss knolls of the Rogart Park, Blackbird and Thrush among the Whin and Gorse, Robins' nests in the hazel woods, also that of the tiny Wren. Stonechat and Yellowhammer nested among old dykes, and of course, there were the House Sparrows, which lived underneath our roan pipe and wakened us up with their chatter.

In Achnagarron Wood we would find a Tawny Owl nesting. We would climb the tree and look in at the round eggs; the Owl hardly moved. There was also a Buzzard nesting nearby. Buzzards were very rare; the gamekeepers were always after them. The Hoodie Crow was another which was much persecuted, along with any other bird or mammal which was a threat to game birds. The Rook colonies in Rovie Farm and the Station firs also came in for severe bombardment at nesting time. Jackdaws seemed not to be so numerous as they are today; Wood Pigeons were probably more so.

Game birds were common. Every croft district had a covey of Grey Partridge; you would find them in the stack yard, picking up on the oat spill. Grouse were also common, especially on the high ground of the parish; people spoke of their corn stooks being black with Grouse. Even the in-bye hills of Pittentrail, Rhemusaig and Pitfure would have a covey or two on them. Their cackle on a morning of frost is a lasting memory.

Every spring was heralded in with the Peewit, its call a welcome to a new season of growth. Around our way, they favoured the marshland of the Rogart Park, nesting on the tussocks, but most hills had a few, especially around the cross-roads, where they were very numerous. Another springtime arrival was the little Black-headed Gull; every loch had them, nesting on the marshy surround and islands. There must have been thousands of them over the entire parish; big flocks would descend whenever the plough turned the first sod of the spring. Now you will find none.

Hedgehogs were common; occasionally you would come on a nest with young. Stoat and Weasel were seen hunting the dry stone dykes for rabbit. Any hill which had Grouse would also have Mountain Hare, a common and popular source of food. Like-wise the Rabbit, which was very numerous, and provided a seasonal income to many crofters.

Rats and Mice were on every croft. When the bottom of the corn stack was taken in, and their final refuge exposed, it was pandemonium time, as dogs were called round to dispatch as many as possible before they escaped.

Corry Burn always had its Brown Trout; nothing too big, but a few would make a nice meal. Eels were in the deep, still pools; some grew pretty big, and it was exciting to find a waterfall when the elvers were making their ways upstream. The favoured time for fishing was after the first spate of the summer, when Sea Trout would also make their way up the burn. There was always quite a good run, but father would say that it was nothing to the run of his day. Salmon were numerous in the Fleet, coming up river from early summer onwards.

Starlings seemed to be very numerous. A man who was very active around the parish at that time, encouraging this species in particular, was an uncle of Sandy Mackenzie at the Post Office. His uncle Robert was a scientist, recording birds and setting up nest boxes all around the parish, especially near the school. He published a paper on the Starling.

Our schoolteacher, Mrs Mackay, was enthused and made us aware of the need to protect birds. She was also an expert in beekeeping, and, as many people had a beehive on the croft, we were quite clued up on the rudiments of this husbandry. At Reidlin, where the thatched house was, they had still the old-style straw skep beehives. The honey was cut out of the topmost skep, into a basin.

We sometimes came across an Adder, especially around Balchlaggan Rock and Reidlin. They were fairly common on the outer hills, and caution was exercised when turning over a lying divot, during peat-cutting. Adders would appear in early spring, with a blink of sunshine, but the best time to find one was when it was basking in the summer sun. The instinct of most people at that time was to kill the creature. Only once did we see a family of young adders; that was below Laide, on the old feal dyke, beside the Corry Burn. They were usually near to water.

# Habitat and species survey, 2005         *Viv Halcrow*

Rogart is a small community of scattered houses and farms. The station, hotel, shop, garage, and more recently-built houses are located along the A839, on the flat ground of Strath Fleet, while most of the farms and older housing lie in hilly ground to the north. The level ground in the Strath is of relatively high agricultural quality. There is evidence of a long history of human settlement: a broch and chambered cairns lie close to the village.

The survey on which this account is based took place on 11$^{th}$ August, 2005. The area covered was centred on the village, extending west to Macdonald Place and east to include Remusaig, south to the forestry edge and north to Creag a'Bhata and Millnafua Bridge, approximately 3 square kilometres in all. The purpose of the survey was to provide a general view of the habitats present in the area and some estimate of the range of species that they might support. The few additional comments in square brackets are based on subsequent survey work.

The geology of the area is quartz-feldspar-granulite, with granite to the west. A major fault line forms Strath Fleet and a small area of Lower Old Red Sandstone lies to the south of this. The soils are fairly acid. At the extreme east of the study area, by the railway line, is an area of glacial moraine deposits.

Many of the houses in the village centre have mature gardens with established trees and flowering shrubs, such as *Buddleja* and privet, which are of value to butterflies and other nectar-feeding insects.

Close to the cross-roads, on the banks of the Garbh-allt or Corry Burn, there are large oak trees, with oak apple galls, aspen and hazel. Aspen is known to support a range of specialised wildlife. Sweet cicely *Myrrhis odorata* occurs on the roadsides; it is an early-flowering umbellifer with aromatic leaves, used by insects emerging in the spring.

The disused sheep market is overgrown and 'weedy', with flowering dockens *Rumex obtusifolius*, nettles *Urtica dioica*, and a variety of grasses. [The posts and rails are covered in lichens and the dense turf harbours a wide variety of invertebrates.] 'Weed' and grass seeds are of value to finches, including chaffinches, siskins and, potentially, goldfinches. Nettles are an important larval food plant for butterflies, including small tortoiseshells, red admirals [and peacocks, which arrived in the area in 2006].

Meadows by the railway have tall herbs such as creeping thistle *Cirsium arvense*, meadowsweet *Filipendula ulmaria*, rosebay willowherb *Chamerion angustifolium* and hogweed *Heracleum sphondylium*; although commonplace, their flowers are an important nectar source for insects and their fruits food for seed-eating birds.

The old mill buildings on the Corry Burn near the station are an excellent nesting site for swallows and house martins, which feed on insects, especially above the river and the cattle-grazed fields to its south. The buildings may also house bats and barn owls. Farm buildings at Davochbeg, West Kinnauld and Remusaig are likely to be put to the same uses [bats are certainly present at Davochbeg].

Lightly sheep-grazed, unimproved pasture north of the river supports a range of flowering plants and grasses. A small white butterfly was seen [the first recorded in the area], molehills occur and grasshoppers are present. Mature trees and large shrubs are a feature of field boundaries north of the river, and the bird cherry present must be a striking feature in the spring.

The large, level fields south of the river are improved pasture, silage or barley, with few mature trees in the field boundaries. The pasture will support the many invertebrates and fungi associated with cattle dung or sheep droppings. Smaller fields opposite West Kinnauld, between the railway and the river, are semi- or unimproved with flowering grasses and have been planted with native broad-leaved trees.

The River Fleet remains fairly natural in character, although flood banks have been built along stretches of it. The water varies in depth, with shingle banks and rocky outcrops providing habitats for a range of aquatic invertebrates [see the SEPA surveys in the section on invertebrates]. A dipper was feeding in shallow water east of Davoch Bridge, alongside a family of grey wagtails: two adults and three juveniles catching insects, likely to include mayflies, stoneflies and alderflies. A chaffinch was feeding in riverside trees. The riverbanks are lined with mature alder, ash, birch, grey willow, and eared willow.

Where grazing animals are excluded and the tree canopy along the river is less dense, there is lush growth of tall herbs including valerian *Valeriana officinalis*, sneezewort *Achillea ptarmica*, knapweed *Centaurea nigra*, ragwort *Senecio jacobaea*, field woundwort *Stachys arvensis*, water avens *Geum rivale*, great woodrush *Luzula sylvatica*, meadowsweet and hogweed. A hedgehog dropping was noted on Davoch Bridge.

South-facing, dry, rocky slopes north of the village centre, to the west of the garage, support lightly-grazed dry heath, dominated by bell heather *Erica cinerea*, gorse *Ulex europaeus*, grasses and scattered birch trees. In places, ragwort, dockens and creeping thistle are frequent. These areas are a rich food source for many insects. Road verges and ditches by Macdonald Place had brambles, raspberry, knapweed, hogweed, tall grasses and perforate St John's-wort *Hypericum perforatum* [new to the area, but possibly a garden escape]. A yellowhammer was heard, and also willow warbler and wren in trees near Heath Cottage.

North of Heath Cottage, an area of dense, tall bracken gives way to dry heath on rocky slopes, with patches of herb-rich grassland, all heavily grazed by rabbits. Bees of a number of different species, including honeybees and buff-coloured bumblebees, were frequent; the holes of mining or solitary bees were evident along open paths. Meadow brown butterflies and grasshoppers were noted. A woodpigeon was seen flying over this area. Areas of gorse have been burned in the past. Rock exposures are frequent, supporting crustose lichens and low-growing plants such as heath pearlwort *Sagina subulata*.

Further east of Heath Cottage, birch woodland occurs, with frequent juniper, and a sparse ground flora, including grasses, wood sage *Teucrium scorodonia*, self-heal *Prunella vulgaris,* wild strawberry *Fragaria vesca* and, rarely, rockrose *Helianthemum nummularium*. The last is a local speciality. Speckled wood and meadow brown butterflies were seen in glades, with hoverflies and other flies feeding on creeping thistle.

Hay is cut in fields north and west of Corry, and west of Millnafua Bridge. Management for hay is less intensive than silage and benefits wildlife, in that artificial fertilisers are not used, and the crop is cut late in the season, allowing flowering plants to set seed.

Alongside the Garbh-allt, where it flows gently [most of the time] south of Millnafua Bridge, mature birch, hazel, gean, bird cherry, rowan, grey and eared willows occur, along with juniper and wild roses. Tall herbs flower by the burn: marsh hawk's-beard *Crepis paludosa*, meadowsweet, valerian, knapweed and hogweed. A common blue butterfly was seen at Millnafua Bridge. Among ruined buildings are nettles, thistles and dockens. Remusaig has small fields, mostly unimproved or semi-improved grassland, with trees along field boundaries, and dry tracks between houses, which may provide habitat for invertebrates such as tiger beetles *Cicindela campestris* [seen elsewhere in 2006]. Outbuildings and ruined buildings are frequent. Birch woodland with juniper occurs on higher slopes, with patches of birch regeneration established on heathy slopes to the east. Along the tarmac side-road to Remusaig there were many toads squashed dead, a large ground beetle *Carabus* sp. on the road, and meadow brown butterflies and grasshoppers on the verges. A grey wagtail was noted on the burn.

The burn bends to flow westwards towards the village centre and, west of a waterfall adjacent to a disused quarry, it is contained within a small, steep-sided gorge for approximately 500m. Here, large oak trees dominate, with aspen, alder, bird cherry, birch, rowan, hazel, blackthorn and juniper frequent. Gnawed holes in fallen hazel nuts indicate the presence of wood mice. The steep sides of the burn make access difficult and grazing is minimal. The ground vegetation is suited to shady conditions, with great woodrush frequent, and climbing ivy *Hedera helix* and honeysuckle *Lonicera periclymenum*, the latter with leaves attacked by leaf miners. A number of large holes among tree roots, and a steep worn path to the burn, indicate the presence of otters, although no spraints were seen. Small waterfalls, pools and boulders create diversity in the course of the burn. Rocks in the burn were uncovered following a dry spell of weather and supported a luxuriant growth of bryophytes. The sheltered, humid nature of the gorge makes it suitable for a range of mosses and liverworts. Meadow brown butterflies were noted on the woodland edge south of the burn, adjacent to open grassland, and large hoverflies were feeding on flowers of cat's-ear *Hypochaeris radicata*.

North of the playing field is a small area of flowering tall herbs, and woodland alongside the burn had blue tit, great tit, willow warbler, siskin and chaffinch. Molehills were noted here. Above and east of the village centre lies a large area of ungrazed (by stock) grassland, heathland and birch woodland (Ard a'Chlachain). The lower slopes are grassland, dominated by sweet vernal-grass *Anthoxanthum odoratum* and Yorkshire-fog *Holcus lanatus*, with flowers such as eyebright *Euphrasia* agg., ribwort plantain *Plantago lanceolata*, red clover *Trifolium pratense* and lesser stitchwort *Stellaria graminea*. Orange-coloured bumblebees and large and small buff-tailed bumblebees were feeding on red clover, while grass-moths, chimney sweeper moths, honey bees, crane flies and other flies were frequent. Meadow pipits and siskin were feeding amongst the vegetation, and a buzzard was hunting overhead.

Further east, on the highest part of the open ground, heavily rabbit-grazed heath vegetation dominates, with heather *Calluna vulgaris*, devil's-bit scabious *Succisa pratensis*, tormentil *Potentilla erecta*, and carnation sedge *Carex panicea*. There are exposed rock slabs with crustose lichens including *Parmelia saxatilis*, and the upright *Cladonia portentosa* and *C. uncialis*. This heath area merges into birch-dominated woodland, with occasional oaks to the east, sloping down towards West Kinnauld. A great spotted woodpecker was heard, in addition to rooks, coal tit and greenfinch, and a large flock of siskins was seen over the woodland. On wetter ground near the old curling pond, willow carr dominates, with birch and alder frequent. Many of the grey willow leaves were covered in a rust fungus, while mite galls were present on alder leaves, and lichens were frequent on the birch trees. A large dumbledor beetle, probably *Geotrupes stercorosus*, was noted, and also wasps *Vespula vulgaris*.

Open water in the old curling pond supported the common back-swimmer *Notonecta glauca* and whirligig beetles *Gyrinus* sp. A range of water plants occurs, including bogbean *Menyanthes trifoliata*, marsh cinquefoil *Potentilla palustris* and bog pondweed *Potamogeton polygonifolius*. The slender sedge *Carex lasiocarpa* occurs at the northern edge of the pond. A scarlet hood waxcap fungus *Hygrocybe coccinea* was noted near the pond edge. Invertebrates were frequent in the area, including black slugs *Arion ater*, Highland darter dragonflies *Sympetrum nigrescens*, many blue-tailed damselflies *Ischnura elegans* and emerald damselflies *Lestes sponsa*, red ants in a nest in a *Sphagnum* hummock, flies and wolf spiders. Young frogs were noted in vegetation around the pond. The carcase of a mallard, which appeared to have been caught on the water and dragged to the pond edge (by a fox?), had a dumbledor beetle, burying beetle *Nicrophorus* sp. and rove beetles feeding on it. Roe deer droppings were found in two locations near the pond.

In summary, **an exceptional range of habitats of value to wildlife occurs within walking distance of the village centre**: mature gardens, derelict buildings, a disused sheep market, unmown road verges, ungrazed tall herb meadows, lightly grazed unimproved and semi-improved grassland, hay meadows, boundary trees, dry heath, birch woodland, oak/mixed woodland, a burn gorge and the river. Current management is essentially wildlife-friendly. Grazing, outwith the improved pasture south of the river, is generally light and roadside verges unmown, allowing plants to flower and set seed. Spraying of herbicides is limited to around the level crossing, and very small areas at house gateways.

It is likely that this range of habitats will support a huge variety of plants, fungi and animals, including the few mentioned above, with species numbering into thousands [as detailed studies are beginning to show].

# Primary School: Wildlife Projects 2005-2007

*Anne Law*

Rogart Primary School is situated on the single track road to Little Rogart, about two kilometres north of the Pittentrail crossroads. It has a wildlife area, which is on the west side of the main building and separate from the main play area. This quiet location makes it an ideal place to observe wildlife.

In the Spring of 2005, we made a pond in the wildlife garden, to provide a habitat for water creatures and plants and also to provide a watering-hole for other animals. A group of children, led by Ian Paterson, the Highland Council Countryside Ranger, dug out the pond and lined it. This was very hard work, as there were huge rocks to take out. Birds drink from and bathe in the pond and we have found pond skaters, beetles and frogs in it. A variety of plants in and around the pond will attract more interest, and we will be pond dipping regularly to see if we have any new residents.

Prior to this, in October 2004, John and Susan Love gave the school a bird box which had a camera attached to it. Susan had once been a pupil at the school. John wired it up for us so that we had a view of the nest interior from our TV in the dining room. In the spring of the following year, a pair of blue tits nested and produced eight eggs. Five chicks survived and we could watch them being fed and cleaned as we had our lunch!

In May, the weather was wet and cold for a few days, so we put some meal worms out on a tray to provide some extra food. Just before the chicks left the nest, the camera stopped working. John discovered that a little animal, possibly a vole, had bitten through the only unprotected part of the cable as it tunnelled round the tree which held the camera. He repaired it using an old bit of hula hoop. The following winter, a blue tit used the nest as a roost at night and was often still asleep when the TV was switched on in the morning! In Spring 2006, a pair of blue tits successfully nested again and five chicks again left the nest.

Earlier that year Ron Graham and Jacqui Heaton from the RSPB had come to the school to catch and ring birds with children. One of the blue tits on the nest had a ring on its leg – so we may have held it in our hands!

In May 2005, we ordered some painted lady butterfly eggs from 'Insectlore'. We put the eggs into little boxes with special food where they hatched into little caterpillars, some growing to 20mm long. Not all of them survived. Just as they were running out of food, they attached themselves to the lid of their box with silk. We carefully moved the cocoons to a butterfly net where they hung for about two weeks. When the cocoons were ready to hatch, they began to wobble about. We watched the emerging butterflies while they pumped blood into their new wings. They fed on ripe fruit for a few days and then we released them in the wildlife area one sunny morning in June. We repeated the whole process again in 2006.

As well as ringing birds in the school grounds, Ron Graham also helped us build bird boxes and a variety of feeders in October 2005. We hung the feeders around the wildlife area and Ron returned in the Spring to put up the bird boxes. We have discovered that the seeds are more popular than the nuts. The children built a bird hide and watched from it using binoculars. We have also tried taking photographs – this is very difficult! We have seen 15 types of birds so far. Some of our favourites are goldfinches, siskins and long-tailed tits.

Ian Paterson brought some rock-rose cuttings to the school in June 2005. We planted the tiny cuttings and then Ian took them back to his greenhouse to bring them on over the summer holidays. In September we planted them in the wildlife garden, hoping that when they grow they will attract the rare Northern Brown Argus butterfly, which occurs elsewhere in the Rogart area.

Each year Councillor Ian Ross has given the Pupil Council a donation to spend on a project of their choice. In 2006 the children spent it on materials to make raised vegetable beds in the playground. We plan to develop this project in 2007.

In September 2005, we applied to take part in a 'Keep Scotland Beautiful' initiative, which aims to encourage schools to work towards achieving an Eco-School Green Flag Award. To be successful, staff and children must demonstrate a whole school commitment to the preservation of habitat and encouragement of biodiversity in the school grounds. We must also continue to learn about conservation of the natural world and how we can help to promote sustainability of the planet's resources. An eco-committee was formed and each child had a particular responsibility. For example, the wildlife team ensures that the bird feeders are full; others keep the school and playground litter free and collect scrap paper for recycling.

We bought two compost bins, one of which we use to collect all sorts of fruit and vegetable waste and the other for leaves from the playground. The 'composters' empty the classroom compost bins almost every day into the big bin. There are flies and slugs in the bin and recently we noticed a white fluffy mould too. We hope to make use of the compost in the future in our vegetable garden.

The assessors from 'Keep Scotland Beautiful' visited the school in September 2006 and presented us with our first Green Flag Award. Our main focus in 2007 will be on energy use.

Mike Ellis, of *Roots to Branches*, has been tending to the willow garden next to the bottom field play area since it was planted in 2002. We have used the willow in art projects and have had two workshops from Mrs Sally Orr on basket weaving. Ian Paterson and Mike also constructed an outdoor willow classroom in the area in December 2006. We applied for funding for this project from 'Awards for All' in July 2006. We hope that this addition, along with hedging which the children will plant in 2007, will attract even more wildlife to our school grounds. Part of the grant will be spent on building a greenhouse in the school grounds made from clear plastic bottles which the children have been collecting. The pupils can potter about in the greenhouse at playtimes bringing on plants for the wildlife area and vegetables for the garden.

We hope that all these activities will continue to be a part of the fabric of school life and that the children will grow up with an understanding of their environment and a desire to protect the natural world.

# WILDLIFE IN DETAIL

## Vertebrates other than birds                           *Ian Evans*

This section covers the **mammals**, **reptiles**, **amphibians** and **fish** of Rogart. We now know that the parish has a representative selection of the species likely to occur but, until the inception of this project, available information was sparse. There are, for example, almost no references to Rogart in the historic texts on the vertebrates of Sutherland, for example that by Harvie-Brown and Buckley (1887).

More than a century later, the *Atlas of mammals in Britain* (Arnold 1993) mapped only ten species for the 10 km square NC70, in which much of Rogart is situated. There were no records for bats, rodents, fox, stoat, weasel or badger. Similarly, the slow-worm is the only reptile or amphibian mapped for NC70 in the recent New Naturalist volume on *Reptiles and Amphibians* (Beebee and Griffiths 2000).

We can do better than that, although information is still very incomplete. The account that follows is offered in the hope that it will provoke those living in and visiting the area to keep their eyes open and put their observations on record. The author would be pleased to receive such observations. Where no date is indicated, records have been made since 1990, mostly during the last five years.

### Mammals

**Hedgehogs** are said by JM to have been common in the 1940s, 'occasionally you would come across a nest with young'. More recently, they have been noted, regrettably usually as road casualties, at Tressady (6903), near Rovie Lodge (7103), Reandoggie (7104), Pittentrail (7201, 7202, 7301) and near Morvich Quarry (7401); VH also found a hedgehog dropping on Davoch Bridge (7201).

**Moles** conveniently advertise their presence by the heaps of earth excavated from their burrow systems. Since they feed on earthworms and other invertebrates, they are found in those parts of the parish where fertile soils support good populations of their prey, along Strath Fleet, Strath Brora and the valleys of their tributaries. Molehills are most obvious in short grassland or arable, especially early in the year, rather less so in woodland, which moles also inhabit. Most of our records are from around the village (7102, 7201, 7202), to its west (Muie 6803 and Reandoggie 7104); up the Garbh-allt (7302, 7204) and, further north, around Knockarthur (7406 and 7506), Bardachan (7008) and West Langwell (6909). They must occur in many other places; we do not know if those include Strath Seilga, on the northern march of the parish.

All three species of **shrews** almost certainly exist in the parish, but the only records to hand are of the **common shrew**. Two skulls of this species were found in a discarded lager can in woodland along the Garbh-allt (7304) and an individual was found dead by SP in Corry Meadow (7202). Water shrews should occur along the rivers and burns and pygmy shrews in the hillier parts of the parish. The former may be distinguished from common shrews by their larger size and the clearly demarcated colour of their coat, black above and silver below, the latter by their tiny size and proportionately longer tails.

Four species of **bats** are now known to occur in the parish. Three **Daubenton's** or **water bats** were picked up on a bat detector, foraging at dusk over the River Fleet (7102) on 16.5.04, and the same number at Davochbeg (7201) on 15.9.04 (DP). This species is sparsely distributed in the north of Scotland, usually in river valleys, where it feeds over the water. The small bats seen in the summer over the built-up areas of the village, and the adjacent River Fleet, are pipistrelles. Both the recently-separated **common** and **soprano pipistrelles** have been identified, by DP and LW, from the pitch of their echo-location systems. Most of the records are of common pipistrelles, but LW had a soprano pipistrelle foraging with them at dusk over the River Fleet on 16.5.04. Pipistrelles inhabit roof spaces and other crevices in buildings, both old and new, which they can access though very small holes. Their hibernation quarters are often separate from those of their brood colonies, where the females congregate to have their young. The late Dr Ian Pennie recorded long-eared bats in Rogart during the 1970s, and it is now known that there are at least two roosts of **brown long-eared bats** in the area, one in a house in the village (7202, see illustration), and another a couple of miles to the east. They have also been recorded foraging at dusk at Davochbeg (7201) and elsewhere.

**Rabbits** are one of the more obvious mammals in the parish, on a par with moles, although they are, historically, relatively recent additions to the fauna of Sutherland. They were reported in the *New Statistical Account* (1834) as occurring 'of late…on the lower ground', and by the 1940s were 'very numerous and provided a seasonal income to many crofters' (JM). They are particularly numerous on lighter soils on the hills around the village (7202, 7301), but also found in woodland along the Garbh-allt (7304), near the loch at Little Rogart (7203), and no doubt elsewhere. Their numbers are periodically knocked back by myxomatosis, and although an undoubted pest, they are an important food source for a number of predators, including stoats and buzzards.

The *NSA* (see above) states that the 'gray **mountain hare** is here common on the higher grounds', and they were still there in the 1940s, when 'any hill which had grouse would also have mountain hare' (JM). They are still to be found, since JM saw one recently at Little Rogart (7304), just moulting out of its white coat, as late in the year as May. Although **brown hares** were said in the *NSA* to be 'exceedingly numerous ... on the lower ground' in the parish, they are no longer found there; MM has not seen one in 30 years. Their numbers appear to have declined recently elsewhere in south-east Sutherland, as at Dornoch (DP).

Rogart does have rodents. **Field voles** are probably the most abundant small mammal in the parish, inhabiting rough grassland, field and woodland margins, young plantations, river banks and areas dominated by rushes, if not too wet. Their runs can easily be found in grass and rush tussocks, and their presence confirmed by piles of bright green droppings. They are an important item in the diet of many predators, including, amongst mammals, weasels, stoats, foxes and pine martens, and, amongst birds, kestrels, tawny owls and buzzards. Their runs were found in tussock grassland in the old sheep market (7201) and at Corry (7202), one was brought in by SP's cat in Gordon Place (7201) and another found dead on Corry Meadow (7301).

**Bank voles** may be distinguished from field voles by the chestnut tinge to their fur and proportionately longer tails. They are more restricted in their habitat preferences, liking some shrub or tree cover. Hazel nuts with a neatly-gnawed opening are a sure sign of their presence and these have been found, together with many more opened by wood mice, under hazels on the banks of the Corry Burn/Garbh-allt (7202, 7304).

**Water voles**, belonging to the glossy, black Highland race, were living along the River Fleet in the 1970s and 1980s, when MM found several dead, in the winter, on the nearby roadside at Davochbeg (7201). SP's cat brought one in to Gordon Place at the end of December 2006, so they are still present in the area. The entrances to their burrows, about the size of a tennis ball and surrounded by a neat 'lawn', should be looked for along the river and also on headwater burns in the area.

The second most abundant mammal in the parish is likely to be the **wood mouse** found, as its name suggests, in woodland, especially where hazel is present, but also in and around houses and farms, and in a variety of other habitats, even tall heather. Hazel nuts opened by wood mice have untidy tooth marks around the opening and are usually much more numerous than those opened by bank voles. They have been found in woodland along the Corry Burn (NC7202) and also up the Garbh-allt (NC7304). One was also brought in by a cat in Gordon Place (NC7201) and another found dead on the road nearby (7301).

**House mice** were probably once widespread in Rogart, especially around farms and crofts, when there was more cultivation and grain storage. JM records that 'rats and mice were on every croft' in the 1940s, and some of the mice are likely to have been house mice. However, they have disappeared during the last half-century or so from most of Sutherland, or so we believe. Their coloring is drabber than that of wood mice, especially on the underside, which is shining white in the latter. Their larger relative, the **brown rat**, is still present in Rogart, but gets a rather bad press, and records are few. DM has killed at least four at Reandoggie (7104), MM says that they occur 'now and then' at Davochbeg (7201), and there were individuals dead on the road at Tressady (6904) and the east end of the village (7301) on 16.10.06.

The **fox** is conspicuously missing from the mammals mapped for the Rogart area in Arnold (1993). It is certainly present, but, again, does not enjoy a good press, especially in sheep country, although its reputation for lamb-killing is almost certainly exaggerated. Foxes feed in fact on a wide range of small animal life, from insects upwards, and are major predators of small rodents. Their scats are left in conspicuous spots as territory markers, and reveal the seasonal bias in their diet, which can also include berries. MM has seen them in Davoch and Rovie woods, a rabbit snarer at Little Rogart reported that some of his catch had been eaten by foxes and/or cats, and a scat was found up the Garbh-allt (7204).

The flourishing fish populations of Straths Fleet and Brora and their tributaries must support a population of **otters**, although inland territories can be quite large. One was watched hunting a rabbit at Torbreck (7003) in the 1990s and two were seen by LW on the River Fleet east of Davoch Bridge (7201) in 1999. Signs of their presence have been found more recently elsewhere on the Fleet (7400) and on the Corry Burn (7202), and their musky spraints should be looked for at burn junctions and on conspicuous rocks in their courses.

Records of the other mustelid carnivores are scarce, though most of the possible candidates are present in the parish. There are records of **pine martens** from the uppermost part of Strath Fleet (6305), Torbreck (7003) and Reandoggie (7104), and they may occur more widely. AC watched a **weasel** from his house in the village (7202), and MM says she quite frequently sees them crossing the road near Morvich Quarry (7401). The large local population of rabbits must surely support a few pairs of **stoats**, but only three reports are to hand. In the 1970s, MM came upon one rolling an egg away from her hen run at Davochbeg (7201) and also saw one stalking rabbits on the rockface near Morvich Quarry (7401). JL and BL have seen one in ermine crossing the road near Remusaig (7302) during the last five years or so.

JM wrote, in 2002, that 'feral **ferrets** are fairly common nowadays', blaming their presence on a somewhat cavalier attitude amongst certain 'estate proprietors', or presumably their employees, towards ones that went to ground during rabbit shoots. They are being reported quite frequently; a large male polecat-ferret was found dead on the road at West Kinnauld (7301) in 2003, and there are other reports, of road casualties or sightings, from Muie (6704), west of Tressady (6903), Torbreck (7003), west of Little Rogart (7203), Millnafua Bridge (7302) and towards Morvich (7400).

There are no reports of **badgers** in the immediate vicinity of Rogart; the nearest known sett is in the Torboll Woods on the southern side of the River Fleet. They may be on the increase in parts of East Sutherland; they are certainly flourishing on what appears to be much less fertile ground in the west and north of the county.

The last local carnivore to consider is the **wild cat**, for which records are mapped in Arnold (1993) for NC60, 70 and 71. The records are all post-1960, and the product of intensive research into the distribution of this species, but further details have not yet been located. It is likely that they relate to the more remote parts of the parish. JM's view (2002) is 'although it is rarely seen, it is there all right'.

Finally, to the largest of our mammals, deer and goats. **Red deer** are widespread in the more remote parts of the parish, but do venture closer to the village at times, especially in the winter. MM has seen them jumping into the eastern end of the Davoch Wood (7300), and JL saw one on a ride in the same general area (7101); further to the west, stalking is offered on the Tressady Estate during much of October. The slightly smaller **sika deer** are said to be present in the vicinity of Rogart, but they keep to the more wooded areas, where their high-pitched whistling call gives them away. They are the progeny of escapes from estates to which they were introduced during the 20$^{th}$ century, and are capable of breeding with the related red deer.

Woodland is also the preferred habitat of the smallest of the three species present, **roe deer**. The *NSA* said of them, 'may always be seen here, but not in great numbers', and the Tressady Estate currently advertises buck stalking in May/June and doe stalking through the winter. Recent sightings or signs are from upper Strath Fleet (6305), Reandoggie (7104), around the old curling pond (7301), and east of the village (7401).

**Feral goats** are present on Marian's Rock and the crags to its west (7401), and also the crags at Morvich, just beyond the parish boundary. The late Dr Ian Pennie (*in* Omand 1982) understood that they had been 'settled there to graze the rock to keep the sheep from difficult places', as was the custom elsewhere in Sutherland, especially on precipitous parts of the west coast.

## Reptiles

All three Scottish reptiles are now known to occur in Rogart, where they show a preference for warm, south-facing slopes with relatively short, undisturbed vegetation and open areas in which to bask. Muirburn, especially when carried out late in the winter, after they have begun to emerge from hibernation, can wipe them out locally.

A male **common lizard**, with a striking orange belly, was found by SP dead on the road at Corry (7202) in 2005 and another was seen amongst the junipers high on the hill above Corry on 8.5.06; SW reports that they are frequently seen in that area. Another was seen beside the mill on the Corry Burn (7201) on 5.6.06. The only recent reports of **slow-worms** are of ones seen by SP at Pitfure West (7003) in 2005, and by NM at Reandoggie (7104) in 2006, but JM reports that they also occurred on The Kerrow above Strath Brora (7609) in the 1970s.

**Adders**, because of their reputation, feature more prominently in the available records of reptiles. It should be said, in their defence, that they tend to keep out of the way of human beings and although their venom is poisonous, the only people to suffer bites are the very foolish or very unfortunate. JM remembers coming across them around Balclaggan Rock and Reidlin in the 1940s, when they were also 'fairly common on the outer hills'. He saw one on a south-facing slope at Balclaggan (7304) in 2002 and reports they used to be very common in the area above Corry (7202), where he last saw them in 2004. They have occurred near Garvoult (7305) since 2000 and, more recently, RS and FW saw a big male at the back of the loch at Muie (6604) in July 2005, and found a female in their polytunnel at Muie (6604) the same month. SP also found one in 2005 on the road at West Kinnauld (7201).

**Amphibians**

We now know that three of the five amphibians which occur in Scotland are to be found in the vicinity of Rogart. All are dependent on water for spawning, but spend most of the rest of their time on land. **Frogs** are the most widespread of the three, but they tend to go unreported because of this. They spawn in quite small water bodies, sometimes manifestly unsuitable ones, such as water-filled ruts in tracks. Our records are few, perhaps because we have not asked the right generation where spawn is to be found! There was spawn at Pitfure (7103) in the early 2000s, adults on Corry Meadow (7301) on 22.5.05, and young ones, probably recently emerged, in vegetation around the old curling pond (7301) on 11.8.05, where an adult was also seen by LB on 21.10.06.

An adult **common toad** was seen in Corry Meadow (7301) on 22.5.05 and VH reported that there were many squashed dead along the side-road to Remusaig (7302) on 11.8.05. These were perhaps also young ones that had recently emerged from a breeding locality. They prefer more substantial bodies of water for spawning and the old curling pond is a likely site. Two were seen at Reandoggie in 2006 and one dead on the road east of the village (7310) on 16.10.06.

The **palmate newt** is the most widespread, in the Highlands, of our three species of newts, although there are isolated records of both smooth and great crested newts to the west of Rogart, around Lairg and the Kyle of Sutherland. It is possible that there has been some mis-reporting of smooth newts in the Highland area by those not so familiar with palmate newts. It is therefore good to be able to report the first definite local sighting of the latter species. Some five pairs of palmate newts were seen displaying in a clear, flowing ditch opposite Davochbeg (7201) on 8.3.05. The males were in full breeding dress, with black, webbed hind feet and characteristic filaments extending beyond the end of their apparently chopped-off tails. MM informs us that this is a regular breeding site. A male was also found at the old curling pond (7301) by DM on 21.10.06 (see illustration).

**Fish**

The north of Scotland, despite its attractions to the fly fisherman, actually has a very restricted range of native fish species. Most celebrated is the **Atlantic salmon**, which certainly runs up the Rivers Fleet and Brora, as also does the migratory form of the brown trout known as the **sea trout**. The account of the parish in the *NSA* (1834) has a substantial paragraph on the spawning runs of these fish, and it is interesting to compare this with a recent description of the attractions of the Tressady Estate, which 'enjoys five miles of single bank fishing on the River Fleet, a small spate river with runs of salmon and sea trout from mid-July to the end of the season, on October 15.' One wonders if numbers have decreased in the intervening 170 years? The stay-at-home form of the **brown trout** is found and fished-for in all local rivers, burns and the handful of hill lochs in the parish; JM remarks that in the 1940s, the Corry Burn, for example, 'always had its brown trout, nothing too big, but a few would make a nice meal.' **Arctic charr** are recorded from NC61, presumably from one of the deeper hill lochs in that area.

JM states that **eels** were to be found in 'the deep still pools' in the Corry Burn in the 1940s and 'that it was exciting to find a waterfall when the elvers were making their way upstream'. MM reports that they still occur in the River Fleet at Davochbeg (7201), and they are recorded by Davies et al. (2004) in the upper parts of both Fleet and Brora catchments (NC60 and 61). Their annual migrations are in the opposite direction to the salmonids, ascending the rivers in spring as elvers and returning some years later as adults on their way to distant spawning grounds in the Sargasso Sea. Any observations of either elvers or adults would be of great interest, since the species seems to be in decline.

The other two species known to occur in the Rogart area are at home both in the sea and freshwater, and remind us that until the building, in

1816, of the causeway at The Mound, the sea reached Rogart. John Henderson stated, for example, in 1812, that 'the tides flow into the river Fleet a quarter of a mile beyond Kinauld, to a place called Balintraid' and also, no doubt as a result of tidal flows, that 'the banks of the river are from six to twelve feet deep along Morvich and Kinauld'. **Three-spined sticklebacks** are recorded from upper Strath Fleet (NC60), but no doubt occur nearer its mouth. The males are particularly splendid in their red and blue breeding dress. The other species is the **flounder**, which does occur just downstream of Davoch Bridge (7201). It can penetrate quite a long way into freshwater, but has to return to coastal waters to breed.

Other species of fish that may occur in the parish are the **minnow**, which has been introduced, as live bait, into many Highland waters, introduced **rainbow** trout, and one or more of the three species of **lampreys**, two of which have been recorded in the catchment of the River Oykel, not far to the south.

Thanks are due to members of the Wildlife Group and to the following, whose initials appear above, and to all those others who have contributed local records to the Highland and national data-bases: Paul Blount, Lawrence Brain, Viv Halcrow, John and Brenda Lunn, Dave and Nina Matthews, Daniel Moran, David Patterson, Susan Priddy, Lyn Wells, Sarah Wheeling and Valerie Wilson. We are also grateful to Murdo Macdonald, who kindly extracted records from the H.B.R.G. data-base.

**References**

Arnold, H. (ed.), 1993. *Atlas of mammals in Britain.* London: H.M.S.O.
Beebee, T. and Griffiths, R., 2000. *Amphibians and reptiles. A natural history of the British herpetofauna.* London: HarperCollins
Davies, C.E. et al., 2004. *Freshwater fishes in Britain, the species and their distribution.* Colchester: Harley Books.
Harvie-Brown J.A. and Buckley, T.E., 1887. *A vertebrate fauna of Sutherland, Caithness and West Cromarty.* Edinburgh: David Douglas.
Henderson, J., 1812. *General View of the Agriculture of the County of Sutherland.* London: G.& W. Nicol.
Macdonald, J., 2002. *Rogart: the story of a Sutherland crofting parish.* Skerray: The Byre.
Pennie, I. *Other wild life* in Omand, D. (ed.), 1982. *The Sutherland Book.* Golspie: The Northern Times.

# Birds

*Roger Smith and Fiona Winstanley*

## Introduction

Rogart's diverse range of habitats, from pasture land and woodland to moorland, with rivers, burns and lochs, supports a huge variety of birdlife, from the tiny Goldcrest to the magnificent Golden Eagle.

Between April 2005 and April 2006, residents of the village carried out a survey of the bird life in their gardens and the surrounding area. The results of that survey are the basis of the list given below; it also includes observations made by the authors around their home at Muie, west of the village, and elsewhere in the parish. We have included a few older records of species otherwise not mentioned.

The order used is that in Alan Vittery's very useful *The Birds of Sutherland* (1997). 104 species are included, but the list is not exhaustive. *The Atlas of Breeding Birds in Britain and Ireland* (1976) includes, for example, the following additional species for NC70: teal, wigeon, common partridge, redshank, dunlin, stock dove, grasshopper warbler and spotted flycatcher, although some of these probably no longer occur in the Rogart area. The authors would appreciate notes of any sightings of species that are missing from the list, or any other observations which add to our knowledge of the local birdlife.

## Systematic list

**Red-throated Diver**: summer breeder, from May onwards, on hill lochs. Often seen flying between the coast and lochs; makes unusual quacking sound in flight. Swims with head angled upwards. Winters at sea.

**Black-throated Diver**: summer breeder, from April onwards, on larger hill lochs, where it both nests and feeds. Can be distinguished from Red-throated Diver by larger size and plumage, which is darker with black-and-white chequering. Swims with head held horizontal.

**Little Grebe**: year-round resident, which can be seen on most hill lochs. It needs thick vegetation for breeding, but will feed on any loch.

**Fulmar**: although not seen in the village itself, nests each year on the face of the Morvich Quarry.

**Cormorant**: although a sea bird, can be seen inland fishing the lochs.

**Grey Heron**: year-round resident; on all waters in and around Rogart, feeding on fish and amphibians.

**Whooper Swan**: winter visitor; occasionally seen, from September onwards, using hill lochs as a stop-over. Distinguished from the Mute Swan by its longer, straighter neck.

**Greylag Goose**: winter visitor; flocks feed regularly though the winter in fields at Davoch and Rovie.

**Mallard**: year-round resident; on all waters in and around the village. The commonest European duck.

**Goldeneye**: winter visitor to lochs in the area, but has bred in south-east Sutherland.

**Red-breasted Merganser**: seen very occasionally on hill lochs.

**Goosander**: seen occasionally feeding on hill lochs, in spring and early summer.

**Red Kite**: occasional visitor. Six birds were introduced into the Black Isle in 1989 and twenty birds each year for the following five years. After seventeen years' successful breeding, the population is now spreading further afield and they can be seen occasionally over Rogart.

**Hen Harrier**: rare resident breeder; not seen in the village itself, but has been in the surrounding area. Sutherland has one-third of the British breeding population and for this reason large parts of the county, including parts of Rogart parish, have been designated as S.S.S.I.s.

**Goshawk**: one seen on the Lairg side of Muie in 2006; resident in Sutherland.

**Sparrowhawk**: resident breeder; nests in woodland interspersed with open countryside, meaning Rogart has the perfect habitat. It can be identified by its flight pattern - rapid wing-beats followed by short glides; often seen around bird tables, as song-birds are its main prey.

**Buzzard**: resident breeder; the most common bird of prey in the Rogart area. Nests in trees and crags around the parish. Although a large bird, its main source of food is carrion, and it can often be seen feeding on road kills.

**Rough-legged Buzzard**: scarce winter visitor; only one sighting on this survey, 18-20.10.06.

**Golden Eagle**: Britain's largest bird of prey; although not resident in the area, there is a known flight-path over the parish. In 2006, one was watched preening in an oak tree on the outskirts of the village; another was seen on 17.1.07 on moorland between Muie and Langwell, feeding on deer carrion.

**Osprey**: summer visitor; a number are known to nest in the area and they can occasionally be seen fishing hill lochs.

**Kestrel**: a widespread resident breeder, which can be seen throughout the parish. Often seen hovering, searching for small mammals. It will hunt everywhere from high hills to pasture land and coastal areas.

**Merlin**: resident breeder; not seen in the village, but may be seen elsewhere in the parish. Smaller than the Kestrel, and can be distinguished in flight by its shorter tail. Feeds on small birds; one was seen attacking sparrows at Muie on 19.6.05.

**Peregrine Falcon**: although not seen in the village, it is a resident breeder in the parish. Britain's fastest bird of prey; it feeds almost entirely on birds, which it catches in a spectacular pursuit flight.

**Red Grouse**: resident breeder, once very common in the parish. In the 1950s, the Tressady Estate was famous for its grouse shoots, up to two thousand brace being shot in a season. Sadly, only a few can now be seen, although there was an increase in numbers in 2006, due to a very good year for heather growth.

**Black Grouse**: resident breeder, but very scarce; a few may occasionally be seen on the moorland. One on 18.1.07 between Muie and Langwell.

**Red-legged Partridge**: introduced breeder; one in garden and adjacent field, The Quest, 22.4.06.

**Pheasant**: resident breeder, originally introduced from Asia for sport. Very common throughout the parish.

**Oystercatcher**: coastal resident, which often comes inland to breed, and may be seen from March onwards throughout the parish. A comical-looking bird, which often nests in strange places, e.g. freshly-dug potato-patches or on top of fence strainers.

**Lapwing or Peewit**: once a common sight in Rogart, but numbers are now declining, possibly due to changes in farming methods. Distinctive and skilled flyers, usually seen in April and May, over ploughed fields and pasture land.

**Snipe**: summer breeder, in small numbers; wintering birds from elsewhere?

**Woodcock**: year-round resident of woodland in and around the village. A difficult bird to spot, as its plumage blends with the woodland floor, but may be seen at dusk along the side of woodland; when disturbed takes off in a zig-zag flight.

**Curlew**: coastal resident, which comes inland to breed; can be seen from April onwards, on damp meadows and moorland around the parish. Has a very distinctive call in flight.

**Greenshank**: occasionally seen, but more often heard, along the River Fleet.

**Common Sandpiper**: summer visitor; can often be seen on and around the River Fleet. It has a distinctive teetering walk.

**Black-headed Gull**: formerly bred ('thousands of them over the entire parish' in the 1940s, JM); now a regular visitor.

**Common Gull**: returned to breed on Loch Muie 2006; otherwise a regular visitor.

**Herring Gull**: regular visitor, but does not breed.

**Great Black-backed Gull**: regular visitor, scavenging dead rabbits and other carrion; does not breed.

**Woodpigeon**: resident; a common sight throughout the parish.

**Collared Dove**: year-round resident, common in the village and nearby.

**Cuckoo**: summer breeder, arriving in late April and leaving in late July. Seen in the village, but noticeably frequent on the surrounding hillsides, due to the numbers of its favourite host, the meadow pipit; it lays one egg in each nest.

**Barn Owl**: occasional sightings within the parish, especially since 2000.

**Tawny Owl**: year-round resident, throughout the parish. Often seen, but even more often heard, particularly during its breeding season in late January and early February.

**Long-eared Owl**: known to breed in the area, but sightings few.

**Short-eared Owl**: commoner in the area than either Barn or Long-eared Owls; no recorded sightings in the village, but seen in outlying areas. In flight, its wings appear very long in comparison to its body and it flies with long, slow, wing-beats, quartering the ground in search of small mammals. In late 2006, one took up residence in a garden at Muie.

**Swift**: once a regular summer visitor to the village; not for a few years.

**Kingfisher**: only one report; not seen during 2005.

**Great Spotted Woodpecker**: resident; seen occasionally in the wooded parts of the village, but more often heard.

**Skylark**: resident breeder; once a common sight everywhere, but now scarce. Seen in 2005, but none reported in 2006.

**Sand Martin**: summer breeder; seen March-September along the banks of the River Fleet.

**Swallow**: summer breeder; arrives late April, leaves early October. Nests in barns and outhouses throughout the parish. The arrival of the swallow is seen to be the start of spring.

**House Martin**: summer breeder; seen in the village from May to September, nests under the eaves of houses.

**Meadow Pipit**: resident breeder; very common on moorland and pasture, forms large flocks in the winter.

**Yellow Wagtail**: two possible sightings in April 2005, none since.

**Grey Wagtail:** year-round resident; seen along the banks of the River Fleet and on fast-flowing burns.

**Pied Wagtail**: year-round resident; a common sight in the village and the surrounding countryside, often near water. At Muie, frequently seen following horses, probably eating insects that have become trapped in the hoof prints.

**Waxwing**: in the garden of The Quest, 22-26.12.96

**Dipper**: year-round resident; a common sight along the River Fleet, and on burns and lochs throughout the parish.

**Wren**: year-round resident; very common throughout the parish, often seen near water.

**Dunnock**: resident breeder; a common sight in village gardens, often underneath bird tables, feeding on dropped seed.

**Robin**: year-round resident; seen in gardens throughout the parish.

**Black Redstart**: scarce passage migrant in Sutherland; one in the garden of The Quest, 13.5.06

**Redstart**: summer breeder; only one recorded sighting in the parish in 2006, a breeding pair at Little Rogart.

**Whinchat**: summer breeder; only one record, at Little Rogart, in summer 2005.

**Stonechat**: year-round resident, found on heaths and moorland throughout the area. Two recorded sightings on the survey: at Little Rogart, April to September 2005, and at Muie, in November 2005.

**Wheatear**: summer breeder; common throughout the parish, in most hilly areas and rocky countryside, but under-recorded.

**Blackbird**: year-round resident; can be seen anywhere in Rogart.

**Fieldfare**: winter visitor; can be seen in large flocks around the whole parish. However, there is a known breeding population in Sutherland, and birds have been sighted in the parish throughout the year.

**Song Thrush**: year-round resident; common throughout the parish.

**Redwing**: mainly a winter visitor, although a few are known to breed in Sutherland. Four sightings in the winter of 2005-2006; six seen at Muie on 28.11.06.

**Mistle Thrush**: year-round resident; common in the village, although few sightings elsewhere in parish.

**Sedge Warbler**: summer breeder; can be seen, or heard, around lochs and along rivers, such as the Fleet, where there are extensive reed beds or other dense vegetation on the banks.

**Whitethroat**: summer breeder; reported in the Pittentrail area. Once common, but now scarce, perhaps due to destruction of nesting habitat.

**Garden Warbler**: summer breeder; only one recorded sighting in the parish, but probable more widespread than this suggests.

**Blackcap**: year-round resident, although uncommon. Two reported in 2005, one at Little Rogart, one at Muie; a female at Muie on 28.10.06.

**Chiffchaff**: scarce summer breeder; only one record, at Muie on 29.5.05.

**Willow Warbler**: summer breeder; can be seen, or more often heard, in and around the village from April to September; very sharp, clear notes.

**Goldcrest**: year round resident; with the Firecrest, the smallest European bird. Can be seen, or heard, in conifer woods throughout the parish. One flew into a window at Muie in May 2005, and was later released.

**Long-tailed Tit**: year-round resident; common throughout the parish. Often seen in large flocks in the winter.

**Coal Tit**: year-round resident; common throughout the parish.

**Blue Tit**: year-round resident; very common garden bird in the parish.

**Great Tit**: year-round resident; commonly seen on garden bird-feeders throughout the parish.

**Tree Creeper**: year-round resident; in mixed or coniferous woodland around the village.

**Jackdaw**: very common year-round resident; seen over the village in large flocks at evening time, performing aerial manoeuvres before roosting.

**Rook**: year-round resident; common, but definitely under-recorded on the survey.

**Carrion Crow**: common year-round resident, seen throughout the parish. It is not a flock bird, and is usually seen alone or in pairs. Differs from the Raven in its smaller body, shorter neck and straight tail.

**Hooded Crow**: common year-round resident; same species as the Carrion Crow, but with characteristic black and grey colouration; intermediates occur.

**Raven**: year-round resident, common throughout the parish. The largest of the crow family, it can be identified by its larger size, stronger bill and wedge-shaped tail.

**Starling**: once a town bird, but in recent years appears to have spread out to more rural areas; now very common throughout the parish. Traditionally a hole-nester, but a few years ago a small flock moved to a garden at Muie, where they build nests in trees and continue to flourish. There has been a noticeably large increase in the village population.

**House Sparrow**: year-round resident; very scarce at one time, but there appears to have been a large influx over the last two years, making it the most commonly seen bird in the garden.

**Tree Sparrow**: year-round resident; a small flock nests at Muie. Possibly under-recorded in the village. Distinguished from the House Sparrow by its chestnut crown and white cheek with black spot.

**Chaffinch**: very common year-round resident throughout the parish. Often becomes very tame, when it will take food from human hands.

**Brambling**: winter visitor; has been known to breed in Britain, but not commonly. One seen at Achork in April 2006.

**Greenfinch**: very common year-round resident; in most village gardens.

**Goldfinch**: year-round resident; seen in gardens throughout parish. Feeds mainly on seed heads and aphids. Plant teasels to attract this brightly-coloured bird to the garden.

**Siskin**: very common year-round resident. From the survey, it appears that birds over-winter around the village and spread out as the breeding season draws near.

**Linnet**: year-round resident; only seen in Little Rogart from April to October. One sighting in the village, where possibly under-recorded.

**Twite**: year-round resident; only seen May to October at Little Rogart.

**Redpoll**: only two sightings, September/October 2005, at Little Rogart.

**Crossbill**: occasional visitor, though breeds not far away; in garden of The Quest, 13.6.98.

**Bullfinch**: year-round resident; has been seen in the village and surrounding area, but only occasionally.

**Snow Bunting**: winter visitor and passage migrant. Only known to breed in a small area of Scotland. In summer, black and white plumage; in winter, brown upper parts, pink to buff under parts. Small flocks at Muie during the winter; seen jumping up and pulling down grass stalks to get at seed-heads in the snow.

**Yellowhammer**: year-round resident in the village, nesting in low vegetation. Call resembles 'a little bit of bread and no cheeeeeeeeese'.

**Reed Bunting**: year-round resident; no sightings in the village, but two at Little Rogart, and numerous ones along burns and lochs in the Muie area. Sings from perch, whilst flicking tail.

We should like to thank the following who completed survey forms or contributed other useful information: Alasdair Coupar, Dave Butterworth, Pat Eaglesfield, Dave Goodins, Peter and Pat Goring, Margaret Irving, Maggie MacLaughlan, Tony Mainwood, Andrew and Helen Munro, Fred Munro, Morven Murray and Shirley Pearson.

# Invertebrates and plant galls

## 1. Introduction                                                Ian Evans

Sutherland's scenic beauty is acknowledged, as is the unique character of some of its wildlife habitats, such as the Flow Country. However, away from such internationally-recognised sites, information on the wildlife of particular areas can be sparse, or non-existent, especially in the less 'popular' groups of invertebrates.

Rogart is just such an area. A quick trawl of the national invertebrate atlases and information more recently posted on the National Biodiversity Network revealed the extent of our ignorance. The tally for some of the groups of larger land invertebrates for the 10 km square NC70, which includes most of the village and area to its north, was meagre: three records of land molluscs (all slugs!), one of spiders, none of woodlice, harvestmen, grasshoppers and ground beetles (one of the most popular families of beetles), and none of the larger moths. As a result of fieldwork carried out during 2006, we are able to improve on this, although we have still only scratched the surface.

There is a little more information available on some freshwater insect groups, such as mayflies, stoneflies and caddisflies, which are of considerable significance in aquatic food chains and interest to the fly fisherman. The presence of these and other groups in local watercourses is used as a biological indicator of water quality, and they are logged by freshwater biologists from SEPA as part of their regular sampling of lochs and watercourses across Scotland.

There were some existing records of dragonflies and damselflies, bumblebees and butterflies. Tony Mainwood has contributed a detailed account of the butterflies, based on local fieldwork over several seasons. Fieldwork during 2005 and 2006 by Philip Entwistle and Stephen Moran has made some inroads into our ignorance about other groups requiring expert skills, such as the plant-bugs, other hymenopterans, flies, and some of various groups responsible for plant galls. We have included the plant galls at the end of this section, because a majority of gall causers are invertebrates (although some are fungi).

However, there is still little or no information about even less 'popular' invertebrate groups, such as freshwater sponges, platyhelminthes (flatworms, flukes and tapeworms), nematodes, earthworms and leeches, mites and ticks, despite the fact that many of these impinge significantly on our lives, the well-being of our pets and livestock and the fertility of our soils.

So, although we have made a start, the results are very patchy. This is perhaps not so surprising when we reflect that some 29,000 species of invertebrates are known to occur in the British Isles as a whole, and their identification, even when they have been located, can be quite challenging. Sutherland, with a human population of less than half this number, does not harbour many resident experts in this huge range of 'creepy-crawlies' and is a long way from relevant centres of expertise.

We have, therefore, put on record what we do know, in some detail where the information allows, if only to draw attention to some of the more obvious gaps and to encourage someone to fill them. Included are records from the useful lists made during events held in and around the village in 2005, with all the other information that has come to light since. In some cases the locality information from outside sources is no more precise than the 10 km square, but this has been included for completeness.

## 2. Invertebrates other than insects *Ian Evans*

**Molluscs.** The only land molluscs mapped for NC70 in the national atlas (Kerney 1999) are three slugs. They are the large black slug *Arion ater*, which is almost ubiquitous in Highland, a smaller relative of woods, fields and gardens, *Arion circumscriptus*, and the pale netted slug *Deroceras reticulatus*, an omnivorous pest in gardens and cropland. Which posed the question; were there no snails in Rogart?

The large black slug is certainly present; a splendid orange colour variety was found at the old sheep market (7201, 5.6.06) and the more normal form near a ruined house up the Garbh-allt (7204, 5.6.06). A dark spotted form of the great grey slug *Limax maximus* was found in woodland alonside the Corry Burn (7302) on 5.6.06. Snails do occur; examples of the garlic snail *Oxychilus alliarius* were found in an overgrown corner of the former sheep mart (7201, 5.6.06) and in the gardens of Davochbeg (7201), where also occur the humbug-striped white-lipped snail *Cepaea hortensis* (5.6.06) and the large garden snail *Helix aspersa*. The last is not usually found far from the coast in most of Scotland and is very thinly distributed in the north. It should be possible to double or triple the number of species of land molluscs that have been found in the Rogart area.

Freshwater molluscs are fairly thin on the ground in northern Scotland and even the commonest, the wandering pond snail *Lymnaea peregra*, had not previously been recorded. It is present in 'The Dam' at Davochbeg (7201, 5.6.06), in a drain beside the River Fleet (7201, 26.9.06) and almost certainly elsewhere. Its close relative, the dwarf pond snail *Lymnaea truncatula*, notorious as the intermediate host of the sheep liver-fluke, is likely to occur. SEPA have recorded members of the families Lymnaeidae (see above), Ancylidae (River Fleet, 7201, 2.11.00; probably the river limpet *Ancylus fluviatilis*) and Hydrobiidae (River Fleet, 7201, 2.11.00 and 30.3.03; probably Jenkin's spire snail *Potamopyrgus antipodarum*).

The one other freshwater mollusc recorded from the Rogart area is in a different league. It is the freshwater pearl mussel *Margaritifera margaritifera*, a magnificent large bivalve, whose shells may reach six inches (150mm) in length. Much reduced in numbers over the last century or so, owing mainly to the destructive effects of pearl fishing, it is now only found in a relatively small number of Highland rivers. Individuals can live to 80 years, and maintenance of the population depends on the unwitting co-operation of migratory salmonid fish, such as sea trout, on whose gills a vital larval stage, the glochidia, hitch a lift. The species now receives full protection, and pearl fishing is totally illegal. There are records from NC61, 70 and 71 (Strath Fleet and Strath Brora), although the pearl mussel may now be extinct where it occurred, until the 1970s, in the River Fleet close to Rogart (7201). At the other end of the size scale are the pea mussels, *Pisidium* species; several should occur in the Rogart area, but their identification is challenging and no-one has looked.

**Woodlice or slaters.** Only four species of woodlice occur commonly away from the seashore in northern Scotland; none were previously recorded from NC70. The former sheep market (7201) yielded all four on 5.6.06: the common shiny woodlouse *Oniscus asellus*, common striped woodlouse *Philoscia muscorum*, common rough woodlouse *Porcellio scaber*, and common pygmy woodlouse *Trichoniscus pusillus*. They differ slightly in their habitat preferences; for example, *O. asellus* can tolerate non-calcareous conditions, and *P. scaber* salty or drier ones.

*T. pusillus* is small, reddish and sometimes overlooked, but can be very abundant in soil; *P. muscorum* is mainly associated with tussocky grassland. All but *P. muscorum* were also found in the garden at Davochbeg (7201, 5.6.06) and *O. asellus* in two other places. All are likely to be widespread in suitable habitats around Rogart.

**Spiders.** At least a hundred species of spiders should occur in the vicinity of Rogart, but only one had previously been recorded, the widespread, often red-banded, *Enoplognatha ovata* (Harvey et al. 2002). Spiders are a fascinating group, often with very specific habitat requirements, occurring from the seashore to mountain tops, but they are poorly recorded in Sutherland. Regrettably, very few can be identified to species level in the field; specimens must be collected, preserved and examined under a low-power microscope. Six species were found in the course of casual collecting, five of them new to NC70; the records are listed below (in alphabetical order of scientific name):

Mesh-webbed spider *Amaurobius similis*: under bark, beside River Fleet (7201), 26.9.06

Comb-footed spider *Enoplognatha ovata*: swept from grasses, Corry Meadow (7202), 12.7.06 (SAM)

Wolf spider *Pardosa amentata*: former sheep mart (7201), 5.6.06; shingle, beside River Fleet (7201), 5.6.06

Wolf spider *Pardosa pullata*: lichen-rich heath, above Corry (7302), 5.6.06

Dysderid spider *Segestria senoculata*: under dead willow bark, beside River Fleet (7201), 26.9.06

House spider *Tegenaria domestica*: in public toilet (7201), 5.6.06 (this widespread species is curiously rare in northern Scotland)

Wolf spiders were seen also seen near the old curling pond by VH on 11.8.05 (possibly *Pirata* sp.). Up the Strath, in NC60, two further species of spiders have been recorded, the heavyweight orb-web spinner *Araneus quadratus* and the local wolf spider *Pardosa lugubris* (= *P. saltans*) which occurs in or near woodland.

**Harvestmen.** These relatives of the spiders, with 'rotund bodies ornamented with little spikes, two eyes perched atop [and] ungainly legs insecurely attached', are otherwise known as harvest-spiders or daddy-long-legs (as, confusingly, are a group of flies). Some twenty-odd species occur in the British Isles, but none has ever been recorded from near Rogart (Hillyard and Sankey 1989). Some were seen in Corry

Meadow (7202) on 22.5.05, and the most likely species, *Mitopus morio*, was confirmed as present there by SAM on 12.7.06. It is common in a variety of both lowland and upland habitats. A second, readily identifiable species, *Nemastoma bimaculatum*, which is small and black, with two white dots, was found in the former sheep market (7201) on 5.6.06.

**Ticks and mites.** Sheep ticks *Ixodes ricinus* were swept by SAM in Corry Meadow (7202, 12.7.06). This is the commonest species and a wide range of mammals, birds and even reptiles, serve as hosts, including man. The only mites that are readily identifiable by the amateur naturalist are those that cause galls, usually on trees, and these are dealt with separately.

**Centipedes and millipedes.** These two groups of myriapods have in common a large number of legs, though not so many as is suggested by their names. They differ in the number on each segment (one pair in the case of centipedes, two in millipedes) and in their life styles: centipedes are rapacious hunters, millipedes plant-feeders (and sometimes agricultural or horticultural pests). One species of **centipede** had been recorded from NC70, *Lithobius crassipes*, the commonest small member of its group 'over large areas of rural Britain...notably in grassland and acid heath' (Barber and Keay 1988). We can now add its equally widepread, but larger, relative *Lithobius forficatus*, one of which was collected in the garden at Davochbeg (7201, 5.6.06).

There were no records of **millipedes** from NC70 (Lee 2006), and, in fact, very few from East Sutherland, but we are now able to report the presence of four species:

*Cylindroiulus brittanicus*: under loose bark, dead willow, River Fleet (7201, 26.9.06)
*Cylindroiulus punctatus*: former sheep mart (7201, 5.6.06); in garden, Davochbeg (7201, 5.6.06); under loose bark, dead willow, River Fleet (7201, 26.9.06)
*Ommatoiulus sabulosus*: under loose bark, dead willow, River Fleet (7201, 26.9.06)
*Tachypodiulus niger*: public toilet (7201, 8.5.06); former sheep mart (7201, 5.6.06).

*C. brittanicus* is often found under bark; it is rarely recorded in the Highlands and never before in Sutherland, perhaps because adult males are required for identification. *C. punctatus* is the commonest and most widespread species in Britain. *O. sabulosus* is usually found on sand dunes or heathland; large numbers may invade houses at times. *T. niger* is common further south, but infrequently recorded in the Highlands. A further three or four species may occur in the Rogart area. We are grateful to Dr Gordon Corbet for the identification of both groups.

### 3. Minor insect orders  Ian Evans

**Grasshoppers and their kin (*Orthoptera*).** The north of Scotland is not a rich area for those interested in this group of insects, with records of only a handful of species. Grasshoppers were noted in several places around the village in the summer of 2005, including Corry Meadow, north of Heath Cottage and along the road to Remusaig. The most likely candidate was the meadow grasshopper *Chorthippus parallelus*, which has vestigial hind wings in both sexes, especially the females. It displays a bewildering array of colour forms, incorporating grey, green, brown, pink and reddish tints in different parts of the body. The adults are active from July through to October, and the males have a modest chirpy song. SAM was able to confirm the presence of this species on Corry Meadow (7201) from a specimen swept from grasses on 12.7.06.

He also swept, from heathy vegetation on the southern side of Corry Meadow (7301), on the same date, a specimen of the mottled grasshopper *Myrmeleotettix maculatus*. This small grasshopper, with clubbed or thickened ends to its antennae, is typical of drier heaths and moorland, and also the more mature parts of coastal dunes.

The only other likely species in the Rogart area is the small, brownish common ground-hopper *Tetrix undulata*, which is found throughout the Highland area in heathland and woodland, but is often overlooked.

**Earwigs (*Dermaptera*).** This is a group with few species in the British Isles as a whole, and fewer still in the north of Scotland. We do know that they occur in Rogart, however, since MM confessed to having the common earwig *Forficula auricularia* in her garden during a Highland-wide survey in 2000. Apart from their trademark pincers, which are curved in the male and straight in the female, they are unusual amongst insects in that the mother stands guard over her eggs and remains with the young for some months after they hatch. This species was also beaten by SAM from juniper on the edge of Corry Meadow (7301, 12.7.06).

**Mayflies, stoneflies and caddisflies (*Ephemeroptera, Plecoptera, Trichoptera*).** These three orders of insects share an aquatic nymphal or larval stage, and an aerial, sometimes short-lived, adult stage. Mayflies and stoneflies have a three stage life history (egg, nymph, adult), as do damselflies and dragonflies. Caddisflies have a four stage life history (egg, larva, pupa, adult), as do butterflies and moths, with which they share some other features. These three groups feature large amongst those regularly sampled in waterbodies across Scotland by the ecologists of the Scottish Environmental Protection Agency and have for that reason been treated together here.

Ian Milne of SEPA has kindly made available to us recent records, for all groups, from four sampling stations in the Rogart area, three in or on the boundaries of the parish and one (the Morvich Burn) just outside. They are:

1. the River Brora at Dalreavoch (7508)
2. the River Fleet at Rogart (7201)
3. the Garbh-allt, just east of the crossroads at Rogart (7202)
4. the Morvich Burn at Morvich (7500)

The records of mayflies, stoneflies and caddisflies from these sampling stations have been tabulated to give some idea of the diversity present (see table on p. 54).

The material collected is, necessarily, only identified to family level. The presence of species from these families and their relative numbers, together with those of other groups of invertebrates, is one component of a scoring system (BMWP), which enables rapid and repeatable assessment of the biological health and pollution levels of the waterbodies. Mayflies, stoneflies and caddisflies score highly, being pollution-intolerant, some families of beetles score in the mid-range and oligochaete worms score at the low end. All four local watercourses score either A1 (Excellent) or A2 (Good).

The five families of **mayflies** listed contain species from a range of types of watercourses, upland and lowland, fast-flowing or slow-flowing. This is generally true of the six families of **stoneflies**, although the Leuctridae mostly inhabit stony burns, and two common genera in the Perlidae, *Perla* and *Dinocras*, have large-bodied nymphs also characteristic of fast-flowing upland waters.

## SEPA records of mayflies, stoneflies and caddisflies from Rogart

| Order and Family | Locality and Date | | | | | | | | | |
|---|---|---|---|---|---|---|---|---|---|---|
| | River Brora | | River Fleet | | | | Morvich Burn | | Garbh-allt | |
| | 14.03.03 | 18.09.03 | 10.04.00 | 02.11.00 | 16.04.03 | 30.09.03 | 16.04.03 | 30.09.03 | 16.04.03 | 30.09.03 |
| **Mayflies** | | | | | | | | | | |
| Baetidae | * | * | * | * | | | * | | * | * |
| Caenidae | | | | | * | * | | | | |
| Ephemeridae | | | | | | * | | | | |
| Heptageniidae | * | * | * | * | | * | | * | * | * |
| Leptophlebidae | | | | | | * | | * | | * |
| **Stoneflies** | | | | | | | | | | |
| Chloroperlidae | | * | * | * | * | | * | * | * | |
| Leuctridae | * | * | * | * | * | * | | * | * | * |
| Nemouridae | * | * | * | * | * | * | | * | * | * |
| Perlidae | * | * | * | * | * | | | | | |
| Perlodidae | * | * | * | * | * | * | | * | * | * |
| Taeniopterygidae | * | * | * | | * | * | | * | | |
| **Caddisflies** | | | | | | | | | | |
| Beraeidae | | | | | | * | | | | |
| Brachycentridae | * | | | | | | | | | |
| Hydropsychidae | * | * | * | * | * | * | | | * | * |
| Hydroptilidae | | * | | | | | | | | |
| Lepistomatidae | * | * | | | * | * | | | | |
| Leptoceridae | * | | | | | | | | | |
| Limnephilidae | | | * | * | | * | | * | | |
| Odontoceridae | | | | | * | * | * | * | * | |
| Philopotamidae | | | | | | | | | * | |
| Polycentropidae | | * | | * | * | * | | | | |
| Psychomyidae | | * | | | | | | * | | |
| Rhyacophilidae | * | * | * | * | * | * | | | * | |
| Sericostomatidae | | * | | * | | * | | | | |

The stonefly *Brachyptera putata* is an UK BAP species, with a range in the United Kingdom more or less restricted to Highland and Grampian. Records supplied by Ian Milne include the following from the Rogart area: nymph, Pitfure West, Strath Fleet (7003), 19.11.01; nymph, River Fleet, Rogart (7201), 19.11.01; adult, upstream of Dalreavoch Bridge, Strath Brora (7508), 27.3.02.

The 13 families of **caddisflies** encompass a wide range of life styles and preferred habitats, some have free-living larvae that deploy silken nets to capture prey, others have fixed cases, and still others are free-ranging in cases made from a variety of vegetable and mineral matter. The majority recorded, perhaps predictably, given where the sampling stations are located, have a preference for fast-flowing water. This is certainly true of the sole British members of two of the families: *Brachycentrus subnubilis* (Brachycentridae) and *Odontocerum albicorne* (Odontoceridae).

For further information on these and all freshwater organisms (animal and plant) we recommend the excellent photo-guide by Fitter and Manuel (1994). References to other organisms listed in the SEPA samples will be found in the appropriate sections.

**Dragonflies and damselflies (*Odonata*).** This colourful group of insects shares with mayflies and stoneflies a three-stage life history, but their free-ranging aquatic nymphs, particularly of the larger species, are amongst the most voracious invertebrate predators in the underwater world, being equipped with a spiny, fanged mask that shoots out from under their heads. Identification of the nymphs requires practice, but the adults of those species known from in the Rogart area are not too difficult to distinguish. Bear in mind, however, that the adults are highly mobile and some are migratory. Under the influence of global warming, new species are establishing themselves in the south of the British Isles and others are moving further north.

We have records from NC70 of six species. Three were noted by VH at the old curling pond (7301) on 11.8.05: the emerald damselfly *Lestes sponsa*, the blue-tailed damselfly *Ischnura elegans*, and the Highland darter *Sympetrum nigrescens*. The blue-tailed damselfly was also seen in Corry Meadow on 24.7.05. PFE and SAM also recorded emerald and blue-tailed damselflies at the curling pond on 12.7.06, together with black darters *Sympetrum danae*. Also recorded from NC70 are the large red damselfly *Pyrrhosoma nymphula* and the common hawker *Aeshna juncea*. All six species are widespread in Highland in a variety of aquatic habitats, although the black darter has a preference for boggy pools and lochans.

Further up the Strath, in NC60, two further species have been noted, the common blue damselfly *Enallagma cyathigerum* and the splendid golden-ringed dragonfly *Cordulegaster boltonii*, largest of the local species; these no doubt also occur in the vicinity of the village. The one remaining species likely to be seen is the four-spotted chaser *Libellula quadrimaculata*, a broad-bodied species of still water habitats, with dark spots on its wings, which has been noted in both NC71 and NC80.

**Lacewings (*Neuroptera*).** There are only two records of this group of insects from the Rogart area. A green lacewing, probably a species of *Chrysopa*, was noted on Corry Meadow on 24.7.05, and SAM beat one of the brown lacewings, *Hemerobius* sp., from oak at the edge of Corry Meadow (7202) on 12.7.06. A handful of other species probably occur, together with the related scorpion-flies. From the stalked eggs laid on plants by the delicate-looking adult lacewings emerge voracious larvae that are the scourge of smaller insects, especially aphids, which is just as well if you are a gardener.

**4. Bugs (*Hemiptera*).** *Stephen Moran*

Some forty species from this group were found in the various habitats sited around Corry Meadow, in two days fieldwork, on 12.7.06 and 21.10.06. The first day focussed on grassy areas; the second on areas around the old curling pond.

Members of the order are characterised by their mouthparts, which are modified into piercing/sucking tubes, used to access nutrients from the juices of plants or other invertebrates or both. In the **planthoppers** (*Hemiptera: Homoptera)*, the wings are of a uniform texture and held like the sides of a tent; in the **true bugs (*Hemiptera: Heteroptera)***, they are harder near their insertion on the thorax and membranous distally.

The **planthoppers** found included the abundant *Philaenus spumarius*, better known in its larval stages as the 'cuckoo-spit insect', for its habit of surrounding itself with toxic foam to ward off parasites. This was present through the grassland habitats, along with *Evacanthus interruptus*, a distinctive black and yellow species, and the lacy winged *Cixius*. Several more species were found on their host trees: *Iassus lanio* on oak and *Aphrophora alni*, *Oncopsis flavicollis* and *O. tristis* on birch. All of these planthoppers are 'vegetarians'.

The **true bugs** filled most of the species list. The birch shield bug *Elasmostethus interstinctus*, and the forest bug *Pentatoma rufipes*, found here on alder, were the bulkiest of the bugs found. The birch shield bug is a plant feeder, enjoying the catkins of its host. The forest bug probably has a mixed diet.

Far from vegetarian are the flower bugs (Cimicidae). These relatives of the bed bug are ardent predators, piercing other invertebrates for their juices and capable of 'biting' humans quite painfully! Oak produced *Anthocoris confusus* and *Temnostethus gracilis*; *A. nemoralis* and *A. nemorum* being found on a wide range of plants, high and low. The three damsel bugs (Nabidae) seen are also predatory, and have praying mantis-like forelegs with which to grip their prey.

The most prolific family of true bugs at Rogart was the Miridae, some of which are host specific, but the rest are not so choosy. Many are found in grassland, sixteen species so far in the case of Corry Meadow. Of these, *Stenodema holsatum* is northern and western in its distribution, fading out in Britain towards the south-east. *Pachytomella parallela* is at its most abundant in Scotland, by streams in mountain meadows. Other bugs show preferences for particular host plants, such as *Orthotylus bilineatus* (aspen) and *Cyllecoris histrionicus* (oak), both pretty well at the northern extremity of their UK range. The most notable Mirid found was *Lygus punctatus*, a bug associated with juniper and young scots pine, although here it was swept from rushes by the old curling pond, where it was no doubt searching out an overwintering site. The species range is centred around the Grampian highlands in the UK and this is probably its most northerly station to date.

**Bugs found on Corry Meadow, 12.7.06 and 21.10.06**

**Planthoppers**

*Cercopidae*
   *Aphrophora alni*: beaten from birch, 12.7, 21.10
   *Evacanthus interruptus*: swept from grasses, curling pond, 12.7
   *Philaenus spumarius*: swept from grasses, 12.7; swept from rushes,
      21.10

*Cicadellidae*
    *Oncopsis flavicollis*: beaten from birch, 12.7
    *Oncopsis* cf. *tristis*: beaten from birch, 12.7

*Cixiidae*
    *Cixius* sp.: swept from grasses, meadow, 12.7

*Iassidae*
    *Iassus lanio*: beaten from oak, 12.7

**True bugs**

*Acanthosomatidae*
    *Elasmostethus interstinctus*: beaten from birch, 12.7, 21.10

*Cimicidae*
    *Anthocoris confusus*: beaten from oak, 12.7
    *Anthocoris nemoralis*: swept from grasses, meadow, 12.7
    *Anthocoris nemorum*; swept, heathland, 12.7
    *Temnostethus gracilis*: beaten from oak, 12.

*Miridae*
    *Asciodema obsoletum*: beaten from broom, 12.7
    *Calocoris norvegicus*: swept from nettle, meadow, 12.7
    *Calocoris roseomaculatus*: swept from grasses, meadow, 12.7
    *Capsus ater*: swept from grasses, 12.7
    *Cyllecoris histrionicus*: beaten from oak, 12.7
    *Heterocordylus genistae*: beaten from broom, 12.7
    *Leptoterna dolobrata* swept from grasses, 12.7
    *Leptoterna ferrugata*: swept from grasses, 12.7
    *Lygocoris contaminatus*: beaten from oak, 12.7; from neighbouring birch?
    *Lygocoris pabulinus*: swept from nettle, 12.7
    *Lygus punctatus*: swept from rushes, curling pond, 21.10 (see above)
    *Lygus rugulipennis*: swept from rushes, curling pond, 21.10
    *Mecomma ambulans*: swept fom grasses, 12.7
    *Orthops* sp.: swept from grasses and hogweed, 12.7 (females only)
    *Orthotylus bilineatus*: immatures, beaten from aspen, 12.7 (see above)
    *Pachytomella parallela*: swept, heathland, 12.7
    *Pithanus maerkeli*: swept, heathland, 12.7
    *Plagiognathus chrysanthemi*: swept from grasses, meadow, 12.7
    *Psallus* sp.: swept from grasses, 12.7 (females only)
    *Stenodema calcaratum*: swept from grasses, meadow, 12.7

*Stenodema holsatum*: swept from grasses, meadow, 12.7
*Teratocoris* sp.: immatures, swept from rushes, curling pond, 12.7
*Trigonotylus ruficornis*: swept from grasses, meadow, 12.7
*Nabicula flavomarginata*: swept from grasses, curling pond, 12.7
*Nabicula limbata*: swept from grasses, curling pond, 12.7
*Nabis* sp.: swept from rushes, curling pond, 21.10 (females only)

**Pentatomidae**
*Pentatoma rufipes*: beaten from alder, curling pond, 12.7

[Waterbugs were not collected in 2006, but one species was found by VH in the old curling-pond on 11.8.05: the common back-swimmer or greater water-boatman *Notonecta glauca*].

## 5. Beetles (*Coleoptera*)                                                Ian Evans

Beetles are probably the most successful insect group in the world, and even the relatively impoverished British fauna can muster some 4000 species, with more being added each year. Regrettably, only a few of the larger species are identifiable in the field. One such is the green tiger beetle *Cicindela campestris*, numbers of which were flying on the upper slopes of the hill above Corry (7202) on 8.5.06. Three-quarters of an inch (18mm.) long, with irridescent green wing-cases, this is one of the most active predators in the beetle world, as both adults and larvae. The latter lodge themselves at the top of holes dug in bare soil, their heads acting as camouflaged trapdoors, grabbing any small creatures that venture too close. Despite their colourful appearance, there was no previous record for the Rogart area.

Other large beetles may appear readily identifiable, but they are often members of small groups of superficially similar species with slightly different habitat preferences, to name which requires both practice and access to the technical literature. Examples are the orange-banded sexton beetles and the iridescent blue-black dung beetles, both being represented by several widespread species in the Highland area. Even the familiar ladybirds show a bewildering range of colour patterns in some of the commonest species. We list below just a handful of many hundreds of species that must occur in the Rogart area. Even amongst these few records, there is one, that of the ground beetle *Amara ovata*, that substantially extends the known range of the species in the British Isles.

***Attelabidae*** **(weevils)**
*Deporaus betulae*: swept from grasses, old curling pond (7301), 12.7.06, SAM

***Cantharidae*** **(soldier beetles)**
*Podabrus alpinus*: on alder, River Fleet (7201), 5.6.06, IME
*Rhagonycha fulva*: swept from hogweed, Corry Meadow (7202), 12.7.06, SAM

***Carabidae*** **(ground beetles)**
*Amara aenea*: garden, Davochbeg (7201), 5.6.06, IME
*Amara ovata*: garden, 9, Gordon Place (7201), 8.06, SP/IME; only second record for north of Scotland
*Agonum dorsale*: under stones, floodbank, River Fleet (7201); garden, Davochbeg (7201), both 5.6.06, IME
*Agonum fuliginosum*: under stones, floodbank, River Fleet (7201), 5.6.06, IME
*Carabus nemoralis*: found dead, near curling pond (7301), 21.10.06, SAM
Green tiger beetle *Cicindela campestris*: flying, on upper slopes of hill above Corry (7202), 8.5.06, IME
*Cychrus caraboides*: in rotten conifer log, Davochbeg (7201), 16.10.06, IME
*Leistus fulvibarbis*: under stones, floodbank, River Fleet (7201), 5.6.06, IME
*Nebria brevicollis*: under stone, near Mill (7201); in shingle beside River Fleet (7201), both 5.6.06, IME
*Pterostichus nigrita* s.s.: garden, Davochbeg (7201), 5.6.06, IME
*Pterostichus strenuus*: garden, Davochbeg (7201), 5.6.06, IME

***Cerambycidae*** **(longhorn beetles)**
*Rhagium bifasciatum*: on flowers of bird cherry, Corry (7202), 5.6.06, IME; garden, 9, Gordon Place (7202), 8.06, SP/IME

***Chrysomelidae*** **(leaf beetles)**
*Chalcoides fulvicornis*: swept from nettle, Corry Meadow (7202), 12.7.06, SAM
*Crepidodera transversa*: swept from grasses, Corry Meadow (7201), 12.7.06, SAM
*Galerucella lineola*: eating holes in alder leaves, River Fleet (7201), 5.6.06, IME
Heather beetle *Lochmaea suturalis*; garden, The Quest (7202), 24.4.05; AC; swarmed around this time in both 2005 and 2006

*Coccinellidae* (**ladybirds**)
Larch ladybird *Aphidecta obliterata*: larvae and adults beaten from juniper, Corry Meadow (7301), 12.7.06, SAM
7-spot ladybird *Coccinella 7-punctata*: on fence post, above Corry (7202), 8.5.06, IME; beaten from birch, old curling pond (7301), 21.10.06, SAM
Orange ladybird *Halyzia 16-guttata*: on aspen, Corry Meadow (7202), 12.7.06, SAM; beaten from birch, old curling pond (7301), 21.10.06, SAM

*Curculionidae* (**weevils**)
*Barynotus* cf. *squamosus*: swept from grasses, Corry Meadow (7202), 12.7.06, SAM
*Polydrusus* sp.: swept from grasses, Corry Meadow (7201), 12.7.06, SAM
*Rhynchaenus quercus*: blotch mines in oak leaves, Corry Meadow (7202), 21.10.06, SAM

*Dascillidae*
*Dascillus cervinus*: beaten from juniper, Corry Meadow (7301), 12.7.06, SAM

*Dytiscidae* (**diving beetles**)
*Dytiscus marginalis*: larva, Rogart (70), 11.6.06, AC

*Elateridae* (**click beetles**)
*Dalopius marginatus*: beaten from juniper, Corry Meadow (7301), 12.7.06, SAM
*Hypnoidus riparius*: former sheep mart (7201); under shingle, River Fleet (7201), both 5.6.06, IME

*Scarabaeidae* (**dung beetles**)
*Geotrupes stercorosus*: dead on road, Corry (7303). 5.6.06, IME; on grasses, old curling pond (7301), 21.10.06, SAM

*Silphidae* (**sexton or carrion beetles**)
*Nicrophorus investigator*: on dog dung, old curling pond (7301), 12.7.06, SAM

*Staphylinidae* (**rove beetles**)
*Staphylinus erythropterus*: swept from grasses, old curling pond (7301), 12.7.06, SAM

In addition, the invertebrates recorded at the SEPA sampling stations on the Rivers Brora (7508) and Fleet (7201), the Garbh-allt (7202) and the

Morvich Burn (7500) (see section above on mayflies, stoneflies and caddisflies) included members of the following families of aquatic beetles: *Dytiscidae*, *Elmidae*, *Gyrinidae* (**whirligigs**), *Hydrophilidae*, *Scirtidae* and *Sphaeriidae*.

## 6. Hymenopterans (*Hymenoptera*)     *Philip Entwistle and Ian Evans*

The Hymenoptera are the largest order of insects in the British Isles, encompassing not only the familiar **bees, wasps and ants**, but also the plant-feeding **sawflies**, a huge host of parasitic forms, including the **ichneumon flies**, and tiny relatives (the **gall wasps** or **cynipids**) which are the causative agents in the formation of some plant galls (which see later).

Until 2006, the only significant records of hymenopterans from Rogart were of **bumblebees**, the result of a Highland-wide survey, a detailed and well-illustrated account of which has recently been published by the Highland Biological Recording Group (Macdonald and Nisbet 2006). Murdo Macdonald has kindly made available detailed records from this survey for the Rogart area. PFE recorded the broken-belted bumblebee *Bombus soroensis* and the heath bumblebee *Bombus jonellus* from NC70 on 25.8.00. Both are late species, the former feeding on a variety of flowers on moorland and on forest edges, the latter on bell heather, ling, and thistles, on heath and moorland. On 29.6.03, MM added two further early species, the white-tailed bumblebee *Bombus lucorum*, a common and ubiquitous species of flower-rich grasslands and gardens, and the garden bumblebee *Bombus hortorum* a lowland species of gardens, grassland and roadside verges, which is fond of the flowers of foxglove.

All the above-mentioned bumblebees have been recorded from further up the Strath, in NC60, with two more: the northern white-tailed bumblebee *Bombus magnus*, which is very similar to *B. lucorum*, and the common carder bee *Bombus pascuorum*, a widespread, long-tongued species that favours flowers with a long corolla-tube, such as clovers, vetches and foxglove. Several other species occur in areas to the east and south of Rogart, including one of the parasitic cuckoo bumblebees, *Bombus bohemicus*, so there is scope for further observations of this group in the Rogart area. Anyone interested is referred to the book mentioned above, which provides an excellent introduction to this group of insects.

The only record of a solitary bee to hand is that of the **mining bee** *Andrena subopaca*, swept from grasses in Corry Meadow (7201) by SAM on 12.7.06. We have reports of **honey bees** *Apis mellifera* and just one of a **wasp**, probably *Vespula vulgaris*, beside the old curling pond, both in 2005. PFE collected a specimen of the **sphecid wasp** *Rhopalum clavipes* near the old curling pond (7301) on 12.7.06. Murdo Macdonald kindly identified **ants** collected by IME on 8.5.06 from two nests in the old sheep market. They were the small black *Formica lemani* and the small red *Myrmica ruginodis*, both probably widespread in the Highland area.

Fieldwork by PFE on Corry Meadow on 12.7.06 provided the first records of **sawflies** from the Rogart area. He found the grass-feeding species *Dolerus aeneus* and *Tenthredosis nassata*, and suggests that further species of *Dolerus* will inevitably be present. Also found there was the very common *Tenthredo arcuata*, which is associated with white clover. The leaves of wild roses had been skeletonised by the larvae of *Endelomyia aethiops*, and also associated with the roses was another sawfly *Allantus basalis*, for which the only other Highland records are from the Spey valley in the period 1914-1944.

### 7. Moths (*Lepidoptera*)

*Philip Entwistle, Stephen Moran, Duncan Williams*

Moth trapping, an enjoyable 'lucky dip' approach to entomology, is now an increasingly popular pastime. It has not yet, apparently, taken Rogart by storm, and no reference has yet been located to the operation of traps locally. However, AC has recorded two species of **macro-moths** in his garden (7202): the Sallow *Xanthia icteritia* (24.4.05) and the Brimstone Moth *Opisthograptis luteolata* (7.6.05); VH had the Chimney Sweeper *Odezia atrata* on Corry Meadow on 11.8.05, and SAM a further four species during fieldwork at Corry Meadow on 12.7.06: Common White Wave *Cabera pusaria*, Clouded Border *Lomaspilis marginata*, Silver-ground Carpet *Xanthorhoe montanata* and the Iron Prominent *Notodonta dromedarius* (larva).

We do have, somewhat unexpectedly, a considerable amount of information about **leaf-mining and other micro-moths** from the vicinity of Rogart. Duncan Williams, from near Lairg, has kindly provided a list of the species he recorded in the area (7201 and 7202, unless otherwise indicated), in October 2003 and October 2005. They have been sorted by host, as follows:

**Alder**: *Stigmella glutinosae, Bucculatrix cidarella, Caloptilia elongella, Phyllonorycter rajella, Phyllonorycter froelichiella, Phylolnorycter kleemanella.*
**Ash**: *Caloptilia syringella*
**Aspen**: *Ectoedemia argyropeza, Stigmella assimilella* (near Blairmore, 7304)
**Birch**: *Ectoedemia occultella, Ectoedemia minimella, Stigmella betulicola, Stigmella luteella, Bucculatrix demaryella, Caloptilia betulicola, Phyllonoryter ulmifoliella, Swammerdamia caesiella*
**Bird's-foot trefoil**: *Coleophora discordella*
**Blackthorn**: *Stigmella plagicolella, Deltaornix torquilella*
**Bog myrtle**: *Coleophora lusciniaepennella* (near Achnagarron, 7305)
**Bramble** and **avens**: *Stigmella aurella*
**Hazel**: *Stigmella floslactella, Stigmella microtheriella, Parornix devoniella*; Nut Leaf Blister Moth *Phyllonorycter coryli, Phyllonorycter nicellii, Argyresthia goedartella*
**Oak**: *Stigmella ruficapitella, Stigmella roborella, Tischeria ekebladella, Phyllonorycter quercifoliella, Phyllonorycter harrisella*
**Rosebay willowherb**: *Mompha raschkiella*
**Rowan**: *Stigmella sorbi, Parornix scoticella, Phyllonorycter sorbi*
**Tormentil**: *Stigmella poterii* (near Achnagarron, 7305).

Many of these species may be identified by the characteristic form of the mines created by their larvae; others have to be bred out and checked from the adults. We are grateful to Duncan for this full list of his finds, which shows what may be found in a relatively small area in a few visits, by someone with specialised knowledge.

These records are complemented by those made by PFE and SAM in and around Corry Meadow, on 12.7.06, and 21.10.06. Their list is arranged by family:

*Coleophoridae*
  *Coleophora alticolella*: larvae/cases on rushes, curling pond (7301), 21.10
  *Coleophora lusciniaepennella*: leaf mines on bog myrtle, curling pond (7301), 21.10

*Gracillariidae*
  *Phyllonorycter heegeriella*: leaf mines on oak, meadow (7201), 21.10
  *Phyllonorycter nicellii*: leaf mines on hazel, meadow (7201), 21.10
  *Phyllonorycter quercifoliella*: leaf mines on oak, meadow (7201), 21.10

*Phyllonorycter rajella*: leaf mines on alder, old curling pond (7301), 21.10

**Heliozelidae**
*Heliozela sericiella*; galls/mines on oak near Community Centre (70), 2006; PFE reports that this is only the fifth record in Highland for this species, which is very little known for Scotland as a whole.

**Nepticulidae**
*Ectoedemia albifasciella*: leaf mines on oak, meadow (7201), 21.10
*Stigmella svenssoni*: leaf mines on oak, village (70), 2006
*Stigmella ruficapitella*: leaf mines on oak, meadow (7201), 21.10
*Stigmella salicis*: leaf mine on willow, old curling pond (7301), 21.10

**Ochsenheimeriidae**
*Ochsenheimeria urella*: beaten from broom, meadow (7202), 12.7

**Pyralidae**
*Chrysoteuchia culmella*: swept from grasses, Corry Meadow (7202), 12.7

**Tischeriidae**
*Tischeria ekebladella*: leaf mines on oak, Tressady (NC70), 2006

**Tortricidae**
*Cydia aurana*: beaten from broom, meadow (7202), 12.7

**Yponomeutidae**
*Argyresthia brockella*: beaten from birch, meadow (7202), 12.7
*Argyresthia goedartella*: beaten from oak, meadow (7202), 12.7

## 8. Butterflies (*Lepidoptera*)          Tony Mainwood

Although the range of butterflies in the north of Scotland is relatively limited, the Rogart area hosts a high proportion of them. Records over the last few years account for 16 species and a further three have probably also occurred, but not been officially recorded.

Two species, **Northern Brown Argus** and **Pearl-bordered Fritillary**, are on the UK Biodiversity Action Plan priority list because they have decreased significantly in Britain in recent years. In addition, **Small Pearl-bordered Fritillary, Small Heath, Large Heath** and **Grayling** are currently being proposed for addition to this list. Other resident species include Common Blue, Small Copper, Small Tortoiseshell,

Green-veined White, Dark Green Fritillary and Meadow Brown. The migrant species Red Admiral and Painted Lady add to the richness of the area, while Speckled Wood, Orange Tip and Peacock are more recent arrivals on the scene.

Rogart is particularly significant for the **Northern Brown Argus**, here at its most northerly locality in Britain. It depends on common rock-rose, *Helianthemum nummularium*, which is the food plant for its caterpillars. The Rogart area has an abundance of this species, which itself extends no further north in Scotland, and so far the Northern Brown Argus has been found in 10 different 1km squares (6604, 6704, 6804, 7104, 7202, 7203, 7204, 7302, 7303, 7401). At least one site is within a few hundred metres of the centre of Rogart and there will certainly be more sites to discover. The **Pearl-bordered Fritillary** and the **Small Pearl-bordered Fritillary** both lay their eggs on the leaves of violets. The Pearl-bordered favours areas with patches of bracken on sheltered south-facing slopes and flies in late May and early June, while the Small Pearl-bordered is more often found in damper areas and flies in late June and early July. The two species can, however occupy the same general area, as they do above the old quarry near Morvich. Again more sites will undoubtedly be found for both of them.

Many of the commoner species are, curiously, less well documented than these national rarities and many people will probably be able to provide more information. Details of any butterfly records (listing species, number, date, place and, ideally, grid reference and habitat) would be welcomed by the Highland Butterfly Recorder, David Barbour, at 125a High Street, Aberlour, Banffshire, AB38 9PB (01340 871850; e-mail dbfis@aol.com).

**A check-list of butterflies from the Rogart area:**

    **Large White**: migrant; recorded from NC70 before 1996
    **Small White**: between railway and river (7202), 11.8.05, VH; Corry
      Meadow (7202), 12.7.06, PFE/SAM
    **Green-veined White**: widespread
    **Orange Tip**: two records in 2006
    **Small Copper**: present but probably under-recorded
    **Northern Brown Argus**: recorded from 10 1km squares (see above),
      with up to 30 at two sites
    **Common Blue**: reasonably common
    **Red Admiral**: common migrant, sometimes numerous
    **Painted Lady**: migrant, recorded in variable numbers

**Small Tortoiseshell**: probably reasonably common
**Peacock**: first recorded in 2006
**Small Pearl-bordered Fritillary**: recorded recently at two sites
**Pearl-bordered Fritillary**; known at two sites, but probably present elsewhere
**Dark Green Fritillary**: some records, but probably under-recorded
**Speckled Wood**: three records (7002, 1996; east of Heath Cottage, 7202, 11.8.05, VH; Davochbeg, 7201, 8/9.06, MM), probably under-recorded
**Grayling**: at least one record
**Meadow Brown**: reasonably widespread
**Small Heath**: reasonably widespread
**Large Heath**: several records, but probably under-recorded

## 9. Flies (*Diptera*) *Philip Entwistle*

Flies are the second largest order of insects in the British Isles, mustering some 6,600 species, but their study, especially of the smaller species, is a specialised interest. Prior to 2006, records from Rogart were few. Two widespread species of hoverfly, *Cheilosia illustrata* and *Melanostoma mellinum*, had been noted on Corry Meadow on 24.7.05. In addition, one of a small and highly distinctive family of thick-headed flies, *Conops quadrifasciata*, had been recorded by J.McKellar at Tressady on 31.7.04.

During visits to Corry Meadow by PFE and SAM on 12.7.06 and 21.10.06, particular attention was paid by PFE to this group, and the list that follows records his finds, by family. The records of the gall-midges (*Cecidomyiidae*) will be found in the section on plant galls.

### *Agromyzidae* (leaf-miners)
*Agromyza alnibetulae*: leaf mines, alder, old curling pond (7301). In the British Isles this fly seems to have been found previously on birch, though known from alder on the European mainland; this requires further investigation.
*Phytomyza sphondylivora*: leaf mines, hogweed, meadow (7202) 12.7
*Phytomyza spondylii*: leaf mines, hogweed, meadow (7201 and 7202), 12.7

### *Asilidae* (robber flies)
*Leptarthrus brevirostris*: swept from grasses, meadow (7201), 12.7
*Leptogaster guttiventris*: swept from grasses, meadow (7201) and old curling pond (7301), 12.7

Robber flies are predatory as adults, lying in wait for passing prey on grass stems, twigs etc. There are few species in Highland and none was previously known north of Inverness. So it was a surprise to encounter these two species. *L, guttiventris* may have an association with base-rich soils; we found it in open grassland. *L. brevirostris*, however, seems to have no special soil preferences, which fits with where it was found, both in the meadow and on the edge of the old curling pond, which is likely to be acidic. *L. brevirostris* has been recorded up to the Great Glen and *L. guttiventris* as far as Inverness, though one text says vaguely 'as far north as Sutherland'. The latter species was also found during 2006 on a window of a house at Spinningdale, in the very south of Sutherland.

### *Calliphoridae* (blow-flies)
*Cynomya mortuorum*: swept from hogweed, meadow (7202)12.7

### *Dolichopodidae* (long-headed flies)
Swept from rushes, by the old curling pond (7301), 12.7: *Dolichopus* cf. *vitripennis*, *Dolichopus discifer*, *Hercostomus* cf. *nigriplantis*, *Hercostomus chetifer*, *Hercostomus metallicus*

### *Empididae*
*Kritempis livida, Euempis (Pachymeria) tessellata*: both swept from grasses, meadow (7201), 12.7

### *Heleomyzidae*
*Suillia notata*: swept from rushes, old curling pond (7301), 21.10

### *Muscidae* (house and stable flies)
*Mesembrina meridionalis*: swept from hogweed, meadow (7202), 12.7

### *Opomyzidae*
*Opomyza germinationis*: swept from grasses, meadow (7201/7202), 12.7

### *Rhagionidae*
*Chrysopilus cristatus*:swept from grasses, old curling pond (7301), 12.7

### *Sciomyzidae* (marsh flies)
*Tetanocera ferruginea*: swept, meadow (7201). 12.7; members of this family live in moist habitats and predate water snails

### *Stratiomyidae* (soldier flies)
*Beris geniculata*: swept from grasses, meadow (7201), 12.7

***Syrphidae*** **(hover or flower flies)**
The following species were swept from grasses, rushes or the flowerheads of umbelliferous plants, in the meadow (7201/7301) on 12.7: *Cheilosia illustrata, Epistrophe grossulariae, Episyrphus balteatus, Eristalis horticola, Eristalis nemorum, Eristalis pertinax, Helophilus pendulus, Leucozona glauca, Melangyna labiatarum, Myathropa florea, Platycheirus angustatus, Platycheirus podagratus, Platycheirus scambus, Sericomyia silentis, Sphaerophoria interrupta, Syritta pipiens, Syrphus torvus, Syrphus vitripennis, Volucella pellucens.*

This family deserves special mention. A modest 17 species were found (108 of the 270 British species reach as far north as the east/west line through Rogart). Particularly significant in the context of grassland is the black and yellow *Sphaerophoria interrupta (=menthastri)*, for which this is one of two most northerly records, apart from an old record from the north coast. It was present in numbers in July; its larvae are probably predatory on aphids on various plants. Most of the rest of the species were encountered at an area of hogweed, *Heracleum sphondylium*, close to the Rogart village entrance to the meadow; its flowers are attractive to many insects. Several other hoverflies we found appear to be at, or close to, their northern limit in the British Isles e.g. *Leucozona glaucia, Melangyna labiatarum, Myathropa florea* and *Volucella pellucens*.

***Tabanidae*** **(clegs)**
*Haematopota pluvialis*: swept, meadow (7201) and curling pond (7301), 12.7

***Tachinidae*** **(parasite flies)**
*Tachina grossa*: swept from hogweed, meadow (7202), 12.7; a huge black, bristly fly, whose larvae are parasitic on fox and northern eggar moth larvae

***Tephritidae***
*Chaetostomella cylindrica*: meadow (7201), 12.7

## 10. Plant galls             *Philip Entwistle, Ian Evans, Stephen Moran*

Plant galls are 'abnormal' growths on plants, induced by a wide range of organisms, including bacteria, fungi, nematodes, mites and insects. These organisms have developed systems for side-lining the mechanisms of normal growth in their hosts, so that they provide tailor-made homes and sheltered feeding for the gall inducers. Galls may be small and strictly annual, like the spangle galls which fall off oak leaves in the autumn, or massively perennial, like the witches' brooms on downy birch, which increase in size for decades until, in sheltered situations, they can be two metres across.

Gall inducers are usually host-specific, to a single species or closely related group of species, such as a section of the willows, but the galls induced by closely related organisms may be markedly different in appearance. This means that many galls can be reliably named from the host and their gross morphology. Just as well in many cases, such as the galls induced by mites, since very few naturalists are capable of identifying gall mites.

Although they divert some of the resources of the host to their own ends, gall formers are not generally seriously harmful to those hosts, and, in their turn, provide homes and hosts for a wide range of inquilines (cohabitees) and parasites, and food for larger organisms such as birds. Some species appear to have developed a symbiotic relationship with their hosts and may now be essential to their well-being. Examples are the nitrogen-fixing bacteria in root nodules in members of the pea family, such as broad beans, and perhaps even the mycorrhizal fungi which are now known to live in partnership with 90% of higher plants. For further information on galls and help in identifying them, we would recommend the AIDGAP key by Redfern and Shirley (2002), published by the Field Studies Council.

Galls are of particular interest to several of the naturalists who have been involved in this project, and we do therefore have the makings, albeit incomplete, of a list for the Rogart area, based on observations made in 2005-2006. They are listed below under their hosts.

**Trees and shrubs**

**Alder**: **fungi** *Taphrina sadebeckii* (7104), *Taphrina tosquinetii* (7201,7202); **mites** *Eriophyes inangulis* (6908, 7301), *Eriophyes laevis* (7104, 7201,7301, 7304)

**Ash**: **mite** *Aceria fraxinivorus* (beside River Fleet at and downstream of Davoch Bridge, 7201, the furthest north yet found in the east of Scotland)

**Aspen**: **fungus** *Taphrina populina* (7202); **mite** *Phyllocoptes populi*

(7201); **gall midges** *Contarinia petioli* (7201), *Lasioptera populnea* (7201)

**Birch, Downy** and **Silver**: witches' broom **fungus** *Taphrina betulina* (7103, 7202, 7300, 7302, only on downy birch); **mites** *Acalitus calycophthirus* (7301), *Cecidophyopsis betulae* (7103)

**Lime**: mite *Eriophyes leiosoma* (Davochbeg, 7201)

**Juniper**: **fungus** *Gymnosporangium clavariiforme* (Corry, 7202, orange protrusions on old wood); **gall midge** *Oligotrophus juniperinus* (7201, 7204)

**Rowan**: mite *Eriophyes sorbi* (7201, 7302)

**Oak**: **gall wasps** *Andricus kollari* (6904, 7201, 7202, 7301), *Andricus lignicola* (7202, 7301), *Cynips divisa* (6904, 70), *Cynips longiventris* (Tressady and Rogart village, 70), *Neuroterus albipes* (Tressady, 6904 and Rogart village, 70), *Neuroterus anthracinus* (6904, 7201), *Neuroterus numismalis* (6904, 7201, 7301), *Neuroterus quercus-baccarum* (Tressady, 6904 and Rogart village, 70), *Trigonaspis megaptera* (Tressady, 6904, 70); **gall midges** *Macrodiplosis dryobia* (7201), *Macrodiplosis volvens* (Tressady and Rogart village, 70). Although oaks are scattered, except in the woodland at Tressady, nevertheless they support 11 of the 30-34 gall formers known in Highland, and more may well be present.

**Sallows and willows**: mite *Aculus laevis* (sallows, 7301); **sawfly** *Pontania proxima* (bean gall, willows, 70); **gall midges** *Iteomyia capreae* (sallows, 7301), *Iteomyia major* (sallows, 7201), *Rabdophaga cinerearum* (sallows, 7303)

**Sycamore**: mite *Aceria cephaloneus* (7201)

## Herbaceous plants

**Common dog-violet**: **rust fungus** *Puccinia violae* (7301)
**Common nettle**: **rust fungus** *Puccinia urticae* (Knockarthur, 7506)
**Creeping thistle**: **rust fungus** *Puccinia punctiformis* (7201)
**Germander speedwell**: **gall midge** *Jaapiella veronicae* (7301)
**Nipplewort**: **rust fungus** *Puccinia lapsanae* (7201)
**Tormentil**: **fungus** *Taphrina potentillae* (7201); **gall wasp** *Xestophanes brevirostris* (curling pond, 7301, 21.10.06); the discovery by SAM of this gall on the stem of tormentil is probably a first for mainland Highland; the only other records that have come to hand are from Raasay, Rhum and South Uist, in the 1930s; W.H.J. Trail's

1870s record of a gall on '*Potentilla tormentillae*' near Aberdeen seems probably to refer to this species, rather than *X. potentillae*.
**Wood sage: rust fungus** *Puccinia annularis* (7202, 7401).


## 11. References

Barber, A.D. and Keay, A.N., 1988. *Provisional Atlas of the Centipedes of the British Isles*. Huntingdon: B.R.C.

Fitter, R. and Manuel, R., 1994. *Collins PhotoGuide. Lakes, Rivers, Streams and Ponds of Britain and North-West Europe*. London: HarperCollins.

Harvey, P.R., Nellist, D. and Telfer, M. (eds.), 2002. *Provisional Atlas of British Spiders (Arachnida, Araneae)*. 2 vols. Huntingdon: B.R.C.

Hillyard, P.D. and Sankey, J.H.P., 1989. *Harvestmen*. London: Linnaean Society.

Kerney, M.,1999. *Atlas of the Land and Freshwater Molluscs of Britain and Ireland*. Colchester: Harley Books.

Lee, P., 2006. *Atlas of the Millipedes (Diplopoda) of Britain and Ireland*. Sofia: Pensoft.

Macdonald, M. and Nisbet, G., 2006. *Highland Bumblebees. Distribution, Ecology and Conservation*. Inverness: H.B.R.G.

Redfern, M. and Shirley, P., 2002. *British Plant Galls*. Preston Montford: Field Studies Council.

Centre of village, from plantation above Davochbeg, looking north-east, July 2006: Valerie Coupar

Wooded course of Corry Burn and Corry Meadow, from above Corry, looking south-east, June 2006: Andrew Coupar

Fungus foray, Corry Meadow, September 2005: John Macdonald

Pupils of Primary School filling bird-feeders, October 2005: Anne Law

Palmate Newt, old curling pond, October 2006: Stephen Moran

Brown Long-eared Bats, August 2004: David Patterson

Common Frog, old curling pond, October 2006: Stephen Moran

Roe Deer: Ken Crossan

Stonechat: Ken Crossan

Buzzard: Ken Crossan

Curlew: Ken Crossan

Wheatear: Ken Crossan

Above left – Northern Brown Argus: Tony Mainwood

Above right – Green Tiger Beetle: Pat Evans

Left – Pearl-bordered Fritillary, underside: Tony Mainwood

Below left – Silk-button Galls on oak leaf: Philip Entwistle

Below right – Oak Leaf Mine, *Stigmella* sp.: Philip Entwistle

Bird Cherry: Ian Evans

Common Rock-rose: Tony Mainwood

Above – Pyramidal Bugle: Ian Evans

Right – Bogbean: Ken Crossan

Moss *Antitrichia curtipendula*:
Gordon Rothero

Moss *Grimmia elatior*:
Gordon Rothero

Group of gill fungi *Tricholoma* sp.: John Macdonald

Above – Lichen habitats, old sheep market, mill beyond, May 2006: Anthony Fletcher

Above – Sycamore beside River Fleet, with macro-lichen *Ramalina fraxinea*, May 2006: Anthony Fletcher

Below – Lichen *Ochrolechia parella*, wall-top, May 2006: Anthony Fletcher

Below – Lichen *Acarospora sinopica*, iron railings by level crossing, May 2006: Anthony Fletcher

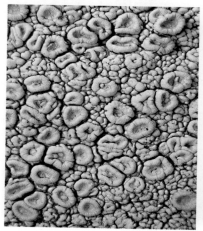

# Microscopic freshwater life: animals  *Ian Evans*

The illustration overleaf shows microscopic animals from a sample collected at the edge of Little Rogart Loch on 29$^{th}$ October 2006. A series of quick pencil sketches was made later, using a high-power microscope, at magnifications of up to 400 times. The author does not pretend to be an artist and some of the organisms were moving at speed, so the sketches give just an impression of the variety of forms present. They are reproduced at a range of scales; the largest were just visible to the naked eye. Identifications are only tentative. Also present in numbers were microscopic algae, of which the desmids in David Williamson's beautiful illustrations (following) are an example.

The selection of animals depicted belong to just five of many groups of microcopic animals present in such habitats.

**1-6. Protozoans.** A huge group of single-celled animals in a bewildering variety of forms, free-living or sessile, solitary or colonial; now placed in several phyla. Most forms are either amoeboid or ciliate.

**7. Gastrotrichs or hairybacks.** A small group of tiny, multi-celled organisms; common, but tend to hide from view; streamlined body, with forked 'tail' processes.

**8-10. Rotifers or wheel animalcules.** A large group of small, multi-celled organisms, with complex contractile bodies, often with two 'wheel organs' at the head end and jaw-like elements in a muscular crop. Free-living or sessile, sometimes with a thickened shell (lorica).

**11. Tardigrades or water bears.** Microscopic arthropods, with a cylindrical body and four pairs of stumpy legs bearing hooked claws. Mostly female; the large eggs are housed in the cast skin. When encysted, they are virtually indestructible. One of the 'characters' of the freshwater world.

**12-13. Crustaceans.** A large group of both microscopic and macro-scopic arthropods, with a segmented body, which in microscopic forms is often more or less enclosed in a carapace, and functionally-specialised jointed limbs (antennae, mouthparts, legs). Second only to the insects in diversity of form and species. Most are aquatic, both freshwater and marine; a few, such as woodlice, can survive on land. Microscopic forms include water fleas (*Cladocera* and *Copepoda*), seed shrimps (*Ostracoda*) and the larval forms of some other groups.

For information on these and other freshwater organisms see the excellent *Collins PhotoGuide to Lakes, Rivers, Streams and Ponds of Britain and North-West Europe* by R. Fitter and R. Manuel (1994), if you can find a secondhand copy.

## Microscopic animals, key to drawings opposite
Pencil sketches of living material; not all to same scale.

### Protozoans
1. Sun animalcule *Actinosphaerium*, x 40: amoeboid, with radiating hair-like pseudopodia
2. Empty test of *Arcella*, x 150: amoeboid, with a brown test (shell) secreted by the animal, which feeds through the hole in the base
3. Empty test of *Difflugia*, x 30: amoeboid, with a vase-shaped test made of sand grains
4. *Naegleria*, x 400: free-swimming amoeboid form that can develop flagellae, as here
5. *Coleps*, x 250: barrel-shaped ciliate, with the body protected by small plates
6. Slipper animalcule *Paramecium*, x 100: fast-moving, text-book ciliate, with mouth situated in a central groove, rotates slowly as it moves

### Gastrotrichs
7. *Chaetonotus*?, x 100: a large genus, with the dorsal surface covered in scales, sometimes produced into spines

### Rotifers
8. and 9. Bdelloid rotifers (without shells): *Philodinavus*? and *Philodina*?, x 100
10. Empty shell of one of the loricate rotifers: *Monostyla*?, x 100

### Tardigrades
11. A female *Macrobiotus*, x 50: with cast skin, not yet detached, containing four eggs

### Crustaceans
12. *Chydorus*, x 30; one of the water fleas or cladocerans
13. A larval crustacean, x 30; no idea what group!

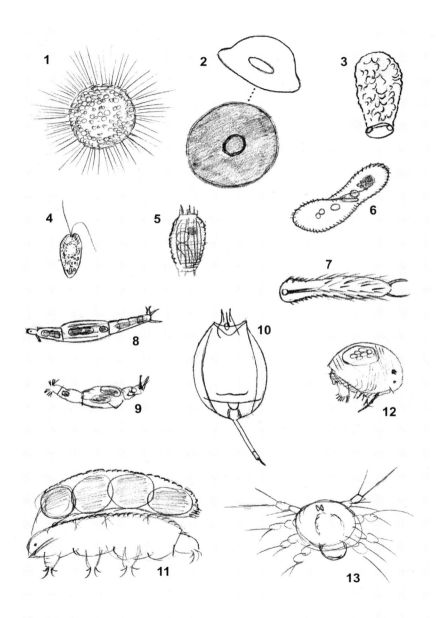

# Microscopic freshwater life: desmids  David Williamson

The desmids illustrated opposite are from a sample collected by Ian Evans at the edge of Little Rogart Loch on 16$^{th}$ October 2006. Desmids are green algae (phylum *Chlorophyta*). World-wide there are around 3000 species, and in the British Isles a total of some 1200 species and varieties.

Desmids live exclusively in fresh water, with the largest numbers and variety of forms found in western parts of the country, from Cornwall, Wales and Cumbria to Western Scotland, including the Western Isles, where they seem to 'prefer' the acidic waters overlying older rock formations. Sutherland has long been regarded as a species-rich area, with many interesting and rare forms. Desmids also occur in the circum-neutral or even alkaline waters of the eastern side of the country, but are here usually found in smaller populations and are generally of a more commonplace or cosmopolitan type.

Desmids are amongst the largest single-celled plants, ranging from 10μm to 1000μm (0.01-1mm), so that the largest can be seen with the naked eye. They exhibit a diversity of shapes unparalleled elsewhere in the plant kingdom, from simple cigar-shaped cylinders or batons to very elongate forms like needles and may be straight, or slightly to extremely curved (lunate). More evolved forms have very complex shapes (e.g. fig. 9), with highly incised margins, patterns of cell wall ornamentation and, sometimes, radiating spines and processes, perhaps used as flotation aids.

Many desmids are planktonic, living in the open waters of ponds and lochs, but they are also found in almost any standing freshwater, such as ditches, cattle troughs, bird baths and gutters, where they live attached,
[continued overleaf]

---

The forms illustrated opposite are as follows: 1. *Closterium costatum* 2. *Closterium angustatum* 3. *Closterium kuetzingii* 4. *Pleurotaenium ehrenbergii* 5. *Closterium dianae* var. *pseudodianae* 6. *Closterium incurvum* 7. *Euastrum oblongum* 8. *Netrium interruptum* 9. *Micrasterias papillifera* 10. *Cosmarium portianum* 11. *Cosmarium margaritiferum* 12. *Staurastrum polymorphum* 13a. *Gonatozygon brebissonii* (with enlarged apex at 13b) 14. *Micrasterias truncata* 15. *Cosmarium brebissonii*.

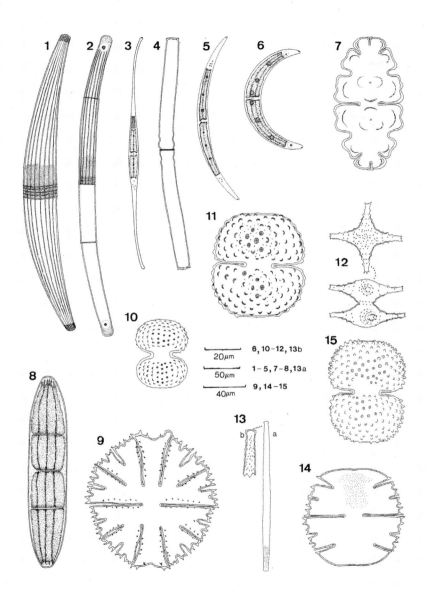

exuding their own 'superglue', to other aquatic plants.

It is little wonder, in view of their attractive and quite remarkable symmetrical shapes, that desmids have been a favourite object of study by microscopists for almost 200 years. It was John Ralfs, a former surgeon turned algologist, who published in 1848 'The British Desmidiae', now accepted internationally as the starting point for modern desmid taxonomy.

The selection of species found in the sample from Little Rogart Loch is typical of a mesotrophic body of water, i.e. one of a moderate nutrient content. Figs. 1-3, 8 and 9 overleaf are of forms more often found in nutrient-poor, acid waters, while the rest could be found in eutrophic (nutrient-rich) waters, more typical of the eastern side of the British Isles.

The sample was collected towards the end of the main summer growing season and, whilst on the whole the mix of taxa is unsurprising, seasonality does play a part, so that samples collected earlier in the year could easily reveal other, even rare, forms.

Other species in the sample, not illustrated, comprised: *Cosmarium tetraophthalmum*, *Closterium baillyanum*, *Closterium cynthia*, *Closterium gracile*, *Closterium intermedium*, *Closterium juncidum*, *Closterium lineatum*, *Closterium lunula*, *Penium cylindrus* and *Penium spirostriolatum*.

# Flowering plants and ferns

Ken Butler, Morven Murray and Viv Halcrow

## Introduction

The area around Rogart has been shamefully neglected by botanists until comparatively recently. Although plant recording in Sutherland as a whole dates back to the 1770s, most of the interest has been in the flora of the coast and the hills, and inland areas have had scant attention. In the case of Rogart, locating older records is made more difficult by some vagueness about the bounds of village and parish.

The main area of the village is centred around Pittentrail Bridge (726200), but the area known locally as Rogart extends north to Little Rogart (7304) and west, up Strath Fleet, to Tressady (6904), Muie (6704) and Acheilidih (6604). The parish, on the other hand, extends over some 105 square miles, from just short of Morvich in Strath Fleet (7400), north-west to the vicinity of Meall an Fhuarain in Ben Armine Forest, north of Lairg (5923). It thus encompasses much of the upper parts of Strath Brora, and is bounded in the north-east by Strath na Seilga, the headwaters of the Black Water. These two areas are remote, difficult of access and rarely visited by botanists.

The traditional unit for botanical recording in Sutherland was the parish and the first general account of the plants of Rogart appears to be a paragraph in *John Anthony's Flora of Sutherland*, published in 1976, four years after his death. Following a brief description of the parish is a list of nine species found 'in the vicinity of Rogart village...all very rare in the county'. They are: *Barbarea vulgaris* [Winter-cress], *Equisetum pratense* [Shady Horsetail], *Helianthemum chamaecistus* [= *H. nummularium*, Common Rock-rose], *Lemna minor* [Common Duckweed], *Lepidium heterophyllum* [Smith's Pepperwort], *Nuphar pumila* [Least Water-lily], *Lythrum portula* [Water-purslane], *Teesdalia nudicaulis* [Shepherd's Cress] and *Vulpia myuros* [Rat's-tail Fescue]. A quick perusal of the body of the *Flora* reveals that the great majority of these, together with the handful of others highlighted for the parish, were recorded, between 1951 and 1960, by the redoubtable Scottish botanist Mary McCallum Webster. The only other name that appears is that of John Anthony himself, who recorded some roses in 1959 and 1960.

Much of Mary McCallum Webster's botanising in the north of Scotland in the 1950s was devoted to gathering records, by 10 km. square, for the *Atlas of the British Flora* (Perring and Walters 1962). The part of Rogart parish with which we are most concerned covers the western two-thirds of the 10km grid square NC70. The total number of species recorded for this square in the *Atlas* was between 250 and 350.

In the 1960s and 1970s, the brilliant naturalist Derek Ratcliffe spent time in East Sutherland, primarily studying birds, but also finding plants. He found three sites for the very rare Rock Cinquefoil *Potentilla rupestris* – one, in 1961, near Bonar Bridge, and two more, in 1976, near Rogart. He also found Don's Twitch *Elymus caninus* var. *donianus* on Torboll Rock, and Rock Whitebeam *Sorbus rupicola* near The Mound.

During 1975-1979, the young Michael Marshall developed a strong interest in plants and wrote good lists of the plants of Rogart. His work was accurate and, knowing the area, he went to places that other botanists had not visited. He went off to University, leaving some useful records.

Since the 1980s, responsibility for recording in all the eight parishes, including Rogart, that make up the botanical Vice-County of East Sutherland (vc. 107) has been that of two Vice-County Recorders, Ken Butler and Morven Murray. Much of their time has been devoted to recording plants, by 10km square, for the *New Atlas of the British Flora* (Preston, Pearman and Dines 2002). The species list that follows includes all plants recorded within the 10km square NC70 up to 2005. It was compiled by them, with added notes on soil preferences, habitats and distribution, by Morven Murray and Viv Halcrow. In total, 450 species are listed, from 86 families.

Morven has this to say, based on the 40 years she has spent in the parish on the family farm at Davochbeg: 'The growing conditions in the parish are very varied, from harsh on the tops of the hills, to milder on the lowland pastures of Strath Fleet. Some sea-shore plants are even found on the roadside verges, thanks to the application of salt and grit in the winter by the Roads Department. This means that the plants found in the area are also very varied.

Although the best time of the year to see many flowers is late June and July, there are treasures to be found in most seasons. Examples are: Moschatel in March at Little Rogart; the succession of Sloe, Gean and Bird Cherry from March to early June; Pyramidal Bugle in late May just east of the Morvich Quarry [see illustration]; and then the late summer flowers, Water Lilies, Water Lobelia, and, of course, several kinds of orchids, which can be seen from early summer through to September.

Spotting and identifying plants adds interest to any walk, and they don't fly off or run away while you're looking at them!'

## List of flowering plants and ferns recorded from the vicinity of Rogart (NC70) up to 2005

The order and nomenclature follow that of the *New Flora of the British Isles* by Clive Stace (2$^{nd}$ Edition, 1997).

[Stonewort, *Nitella flexilis*: The Dam, Davochbeg (7201), 5.6.06, IME, conf. E. Everiss; included here because stoneworts, although algae, are often recorded by botanists working on flowering plants.]

### Clubmoss family, *Lycopodiaceae*
Fir Clubmoss, *Huperzia selago*: acid, nutrient-poor sandy or peaty soils
Stag's Horn Clubmoss, *Lycopodium clavatum*: base-rich soils or in more acidic heaths

### Lesser Clubmoss family, *Selaginellaceae*
Lesser Clubmoss, *Selaginella selaginoides*: damp, base-rich habitats where there is little competition

### Horsetail family, *Equisetaceae*
Water Horsetail, *Equisetum fluviatile*: ditches, ponds and lochs
Field Horsetail, *Equisetum arvense*: riverbanks, roadsides, railways, paths and waste ground
Wood Horsetail, *Equisetum sylvaticum*: deep, peaty, often acidic and damp soils
Marsh Horsetail, *Equisetum palustre*: marshes, damp pastures, ditches and flushes

### Adder's-tongue family, *Ophioglossaceae*
Moonwort, *Botrychium lunaria*: well-drained, base-rich soils

### Polypody family, *Polypodiaceae*
Polypody, *Polypodium vulgare*: dry stane dykes and rock outcrops
Intermediate Polypody, *Polypodium interjectum*: mortared dykes

### Bracken family, *Dennstaedtiaceae*
Bracken, *Pteridium aquilinum*: most vigorous on deep loam

**Marsh-fern family,** *Thelypteridaceae*
Beech Fern, *Phegopteris connectilis*: woodland or amongst rocks, in upland areas
Lemon-scented Fern, *Oreopteris limbosperma*: poorly-drained soils, in upland areas

**Spleenwort family,** *Aspleniaceae*
Black Spleenwort, *Asplenium adiantum-nigrum*: basic crags
Maidenhair Spleenwort, *Asplenium trichomanes*: crags
Wall-rue, *Asplenium ruta-muraria*: basic rocks

**Lady-fern family,** *Woodsiaceae*
Lady-fern, *Athyrium filix-femina*: damp woodland and rocky places
Oak Fern, *Gymnocarpium dryopteris*: woodland and ravines, often with Beech Fern
Brittle Bladder-fern, *Cystopteris fragilis*: damp, shaded, rocky places

**Buckler-fern family,** *Dryopteridaceae*
Male Fern, *Dryopteris filix-mas*: light, well-drained soils
Scaly Male-fern, *Dryopteris affinis*: rocky places
Broad Buckler-fern, *Dryopteris dilatata*: widespread in damp places

**Hard-fern family,** *Blechnaceae*
Hard-fern, *Blechnum spicant*: peaty banks

**Pine family,** *Pinaceae*
Sitka Spruce, *Picea sitchensis*: plantations, self-seeded elsewhere
European Larch, *Larix decidua*: plantations; self-seeded elsewhere
Scots Pine, *Pinus sylvestris*: native and in plantations

**Juniper family,** *Cupressaceae*
Juniper, *Juniperus communis communis*; the upright form, here forming a distinctive shrub layer in birch woodland, also on heaths and crags; Nationally Scarce, but locally abundant

**Water-lily family,** *Nymphaeaceae*
White Water-lily, *Nymphaea alba*: lochs and bog pools
Least Water-lily, *Nuphar pumila*: Nationally Scarce; lochan at Little Rogart and others towards Lairg, often with quaking edges, so beware!

**Buttercup family, *Ranunculaceae***
Marsh Marigold, *Caltha palustris*: wet, base-rich habitats
Globeflower, *Trollius europaeus*: ditches, riverbanks and hay meadows; may be overlooked when not in flower
Wood Anemone, *Anemone nemorosa*: bracken-covered hillsides and woodland
Meadow Buttercup, *Ranunculus acris*: damp, neutral soils
Creeping Buttercup, *Ranunculus repens*: damp, nutrient-rich soils
Bulbous Buttercup, *Ranunculus bulbosus*: well-drained soils, rare locally
Lesser Spearwort, *Ranunculus flammula*: wet places
Lesser Celandine, *Ranunculus ficaria*: damp, shady places on loamy soils
Ivy-leaved Crowfoot, *Ranunculus hederaceus*: edges of small waterbodies
Pond Water-crowfoot, *Ranunculus peltatus*: puddles; in the Park, below the School; northernmost British record

**Poppy family, *Papaveraceae***
Long-headed Poppy, *Papaver dubium*: disturbed ground
Welsh Poppy, *Meconopsis cambrica*: introduced, waste places

**Fumitory family, *Fumariaceae***
Climbing Corydalis, *Ceratocapnos claviculata*: well-drained soils; near the northern British limit
Tall Ramping-fumitory, *Fumaria bastardii*: well-drained acidic soils; northernmost British record on the east coast
Common Fumitory, *Fumaria officinalis*: arable fields and waste ground

**Elm family, *Ulmaceae***
Wych Elm, *Ulmus glabra*: base-rich soils in wet, rocky woodland

**Nettle family, *Urticaceae***
Common Nettle, *Urtica dioica*: damp, nutrient-rich soils
Small Nettle, *Urtica urens*: arable fields and waste ground; easy to overlook; a nettle is a nettle!

**Bog-myrtle family, *Myricaceae***
Bog-myrtle, *Myrica gale*: base-poor bogs and moorland

**Beech family, *Fagaceae***
Beech, *Fagus sylvatica*: native further south; here planted
Sessile Oak, *Quercus petraea*: well-drained soils, the commoner species locally
Pedunculate Oak, *Quercus robur*: heavier soils

**Birch family, *Betulaceae***
Silver Birch, *Betula pendula*: light, well-drained soils, often in even-aged stands
Downy Birch, *Betula pubescens*: wetter, peatier soils
Alder, *Alnus glutinosa*: beside rivers and in bogs
Hazel, *Corylus avellana*: old woodland, on moist, base-rich soils

**Goosefoot family, *Chenopodiaceae***
Fat-hen, *Chenopodium album*: disturbed, nutrient-rich habitats
Common Orache, *Atriplex patula*: disturbed, nutrient-rich habitats

**Blinks family, *Portulacaceae***
Blinks, *Montia fontana*: wet places, often amongst mosses

**Pink family, *Caryophyllaceae***
Thyme-leaved Sandwort, *Arenaria serpyllifolia*: shallow, dry soils
Three-nerved Sandwort, *Moehringia trinervia*: open, moist, shaded ground; here close to its northern British limit
Common Chickweed, *Stellaria media*: disturbed, nutrient-rich habitats
Lesser Chickweed, *Stellaria pallida*: light soils; here close to its northern British limit
Greater Stitchwort, *Stellaria holostea*: woods and shady roadsides
Lesser Stitchwort, *Stellaria graminea*: damp grassland, roadside verges; verges; the star-like flowers are very noticeable in early summer
Bog Stitchwort, *Stellaria uliginosa*: marshes and bogs
Field Mouse-ear, *Cerastium arvense* dry, sandy soils; here at its northern British limit on the east coast
Common Mouse-ear, *Cerastium fontanum*: wide range of fertile habitats
Sticky Mouse-ear, *Cerastium glomeratum*: disturbed ground
Knotted Pearlwort, *Sagina nodosa*: base-rich habitats
Heath Pearlwort, *Sagina subulata*: sandy and gravelly places
Procumbent Pearlwort, *Sagina procumbens*: wide range of habitats
Annual Pearlwort, *Sagina apetala*: dry, gravelly places
Annual Knawel, *Scleranthus annuus*: dry places; here at its northern British limit
Corn Spurrey, *Spergula arvensis*: arable fields and waste ground; known locally as Yarr, so sticky that it used to clog the blade of the mower
Greater Sea-spurrey, *Spergularia media*: on salt-treated roadsides
Sand Spurrey, *Spergularia rubra*: open habitats; tolerant of trampling
Ragged Robin, *Lychnis flos-cuculi*: damp places
White Campion, *Silene latifolia*: deep, well-drained soils
Red Campion, *Silene dioica*: lightly-shaded woodland, roadsides

**Knotweed family,** *Polygonaceae*
[Common Bistort, *Persicaria bistorta*: not in 2002 *Atlas*]
Alpine Bistort, *Persicaria vivipara*: upland base-rich substrates; reproduces by bulbils at the base of the inflorescence
Redshank, *Persicaria maculosa*: weed of cultivation
Knotgrass, *Polygonum aviculare* s.s.: roadsides and other disturbed habitats
Black Bindweed, *Fallopia convolvulus*: weed of cultivation
Sheep's Sorrel, *Rumex acetosella*: dry soils in open situations
Common Sorrel, *Rumex acetosa*: meadows and pastures
Northern Dock, *Rumex longifolius*: roadsides and other disturbed ground; a northern species
Curled Dock, *Rumex crispus*: disturbed ground
Broad-leaved Dock, *Rumex obtusifolius*: disturbed and cultivated ground

**Thrift family,** *Plumbaginaceae*
Thrift, *Armeria maritima*: coastal and montane; only on the highest hills locally

**St John's-wort family,** *Clusiaceae*
Perforate St John's-wort, *Hypericum perforatum*: roadside by Macdonald Place (7202), 11.8.05, VH; new to 70; garden escape?
Slender St John's-wort, *Hypericum pulchrum*: woods and heaths

**Lime family,** *Tiliaceae*
Lime, *Tilia x vulgaris*: introduced; often planted along roadsides

**Mallow family,** *Malvaceae*
Musk Mallow, *Malva moschata*: introduced locally

**Sundew family,** *Droseraceae*
Round-leaved Sundew, *Drosera rotundifolia*: insectivorous; wet heaths, flushes and bogs
Great Sundew, *Drosera longifolia*: insectivorous; loch shores and the wetter parts of bogs

**Rock-rose family,** *Cistaceae*
Common Rock-rose, *Helianthemum nummularium*: short, dry, calcareous grassland, here close to its northern British limit; foodplant for caterpillars of the Northern Brown Argus butterfly

**Violet family,** *Violaceae*
Hairy Violet, *Viola hirta*: introduced this far north
Common Dog Violet, *Viola riviniana*: grassland and woodland
Marsh Violet, *Viola palustris*: bogs, wet heaths and marshes
Field Pansy, *Viola arvensis*: weed of cultivation

**Willow family,** *Salicaceae*
White Poplar, *Populus alba*: introduced; often planted along roadsides
Aspen, *Populus tremula*: suckering tree forming thickets of cloned
 'individuals'; rocky habitats, woods and beside rivers
Crack Willow, *Salix fragilis*: introduced; often pollarded
Osier, *Salix viminalis*: introduced; formerly coppiced or pollarded for use
 in basketry
Goat Willow, *Salix caprea*: usually in woodland
Grey Willow, *Salix cinerea*: wet places
Eared Willow, *Salix aurita*: heathland and along watercourses
Tea-leaved Willow, *Salix phylicifolia*: damp, base-rich soils; northern
 species
Creeping Willow, *Salix repens*: silver-leaved, smallest willow locally

**Cabbage family,** *Brassicaceae*
Thale Cress, *Arabidopsis thaliana*: weed of open habitats
Treacle-mustard, *Erysimum cheiranthoides*; introduced; waste ground
Winter-cress, *Barbarea vulgaris*: damp, disturbed ground; rare locally
American Winter-cress, *Barbarea verna*: introduced; sometimes grown
 as a substitute for water-cress
Cuckooflower, *Cardamine pratensis*: woods and wet meadows
Wavy Bitter-cress, *Cardamine flexuosa*: base-rich soils in shady places
Hairy Bitter-cress, *Cardamine hirsuta*: weed of cultivation; also on rocks
 and by streams
Common Whitlowgrass, *Erophila verna*: dry, open ground
Shepherd's Purse, *Capsella bursa-pastoris*: weed of cultivation
Shepherd's Cress *Teesdalia nudicaulis*: on sands and gravels; Tressady,
 1951, MMcCW; here at its northern limit in the British Isles
Field Penny-cress, *Thlaspi arvense*: weed of cultivation
Smith's Pepperwort, *Lepidium heterophyllum*: pre-1970 record
Rape, *Brassica napus*: escape from cultivation
Charlock, *Sinapis arvensis*: weed of cultivation; less common now
Wild Radish, *Raphanus raphanistrum*: arable fields and waste ground

**Crowberry family,** *Empetraceae*
Crowberry, *Empetrum nigrum*: heathland, mires and rock outcrops

**Heather family,** *Ericaceae*
Rhododendron, *Rhododendron ponticum*: introduced and invasive
Bearberry, *Arctostaphylos uva-ursi*: upland heaths
Heather, *Calluna vulgaris*: often dominant on moorland
Cross-leaved Heath, *Erica tetralix*: mires and wet heaths
Bell Heather, *Erica cinerea*: well-drained slopes and rock outcrops
Cowberry, *Vaccinium vitis-idaea*: heaths and moorland, acid woodland and hummocks in mires
Blaeberry, *Vaccinium myrtillus*: well-drained moorland, acid woodland

**Wintergreen family,** *Pyrolaceae*
Common Wintergreen, *Pyrola minor*: valley of Garbh-allt; uncommon locally
Intermediate Wintergreen, *Pyrola media*: Nationally Scarce; woods and heaths; a few plants locally, all in a relatively small area

**Primrose family,** *Primulaceae*
Primrose, *Primula vulgaris*: woodland, grassland and rocky places
Yellow Pimpernel, *Lysimachia nemorum*: woodland and other shady places
Creeping-jenny, *Lysimachia nummularia*: introduced locally
Dotted Loosestrife, *Lysimachia punctata*: escape from cultivation
Chickweed Wintergreen, *Trientalis europaea*: moorland and under birches or bracken; an attractive northern species
Chaffweed, *Anagallis minima*: usually coastal in northern Scotland; one of the northernmost records on the east coast

**Gooseberry family,** *Grossulariaceae*
Red Currant, *Ribes rubrum*: woodland; scarce in the Highlands
Gooseberry, *Ribes uva-crispa*: introduced; spread by birds

**Stonecrop family,** *Crassulaceae*
Biting Stonecrop, *Sedum acre*: rock outcrops

**Saxifrage family,** *Saxifragaceae*
Yellow Saxifrage, *Saxifraga aizoides*: stony, base-rich flushes and around lochans; the Highlands are its stronghold in Britain
Meadow Saxifrage, *Saxifraga granulata*: on Marian's Rock; here at its northern limit in Britain

Opposite-leaved Golden Saxifrage, *Chrysosplenium oppositifolium*: wet, shaded places

**Rose family, *Rosaceae***
Meadowsweet, *Filipendula ulmaria*: damp, fertile ground
Cloudberry, *Rubus chamaemorus*: wet, base-poor peats in blanket mire, usually above 600m altitude; locally only in the highest areas
Stone Bramble, *Rubus saxatilis*: basic soils, crags and other rocky places
Raspberry, *Rubus idaeus*: dry ground in a range of habitats
Bramble, *Rubus fruticosus* agg.: dry, sunny places
Bramble, *Rubus latifolius*: a northern species
Marsh Cinquefoil, *Potentilla palustris*: mires and wet grassland
Silverweed, *Potentilla anserina*: open habitats, dry or damp
Rock Cinquefoil, *Potentilla rupestris*: base-enriched soils on crags; one of only four localities in the British Isles
Tormentil, *Potentilla erecta*: common on acidic soils
Barren strawberry, *Potentilla sterilis*: infertile, dry soils
Wild Strawberry, *Fragaria vesca*: dry, stony soils
Water Avens, *Geum rivale*: along watercourses
Wood Avens, *Geum urbanum*: damp woodland
Lady's Mantle, *Alchemilla vulgaris*: grassland
Lady's Mantle, *Alchemilla filicaulis*: neutral grassland
Lady's Mantle, *Alchemilla glabra*: grassland and other damp places
Lady's Mantle, *Alchemilla mollis*: frequent garden escape
Parsley Piert, *Aphanes arvensis*: arable fields and open grassland
Slender Parsley Piert, *Aphanes australis*: sandy or gravelly places
Burnet Rose, *Rosa pimpinellifolia*: usually coastal, but here inland in scrub and on cliffs
Japanese Rose, *Rosa rugosa*: frequent garden escape
Dog Rose. *Rosa canina*: wide range of habitats
Glaucous Dog-rose, *Rosa caesia* ssp. *glauca*: well-drained habitats
Sherard's Downy-rose, *Rosa sherardii*: scrub, flowers often deep pink
Blackthorn, *Prunus spinosa*: near to its northern British limit but here frequent; good for sloe gin!
Wild Cherry/Gean, *Prunus avium*: fertile woodlands and ravines
Bird Cherry, *Prunus padus*: moist woodland and shaded rocky places; northerly in Britain, locally common and very attractive in May/June
Rowan, *Sorbus aucuparia*: woods and rocky places
Swedish Whitebeam, *Sorbus intermedia*: introduced, frequently planted
Whitebeam, *Sorbus aria*: native further south, here introduced
Rock Whitebeam, *Sorbus rupicola*: Nationally Scarce; Marian's Rock, DAR, 1976

Himalayan Cotoneaster, *Cotoneaster simonsii*: introduced, bird-sown and invasive
Hawthorn, *Crataegus monogyna*: scrub and open woodland

**Pea family,** *Fabaceae*
Kidney Vetch, *Anthyllis vulneraria*: calcareous soils, mainly on outcrops
Bird's-foot Trefoil, *Lotus corniculatus*: well-drained soils
Tufted Vetch, *Vicia cracca*: roadsides and woodland edges
Hairy Tare, *Vicia hirsuta*: disturbed ground; here close to its northern British limit
Bush Vetch, *Vicia sepium*: roadsides and woodland edges
Common Vetch, *Vicia sativa*: dry, sandy places
Bitter-vetch, *Lathyrus linifolius*: grassland, heath, open woodland
Meadow Vetchling, *Lathyrus pratensis*: roadsides and pastures
White Clover, *Trifolium repens*: grasslands of all kinds
Lesser Trefoil, *Trifolium dubium*: roadsides and other places
Red Clover, *Trifolium pratense*: grasslands
Broom, *Cytisus scoparius*: sandy soils, can be invasive
Petty Whin, *Genista anglica*: heathland; just outside the parish, on The Mound (7798), JKB, 1985; here at its northern limit in Britain
Whin/Gorse, *Ulex europaeus*: undergrazed pastures and disturbed ground

**Water-milfoil family,** *Haloragaceae*
Alternate Water-milfoil, *Myriophylum alterniflorum*: fast-flowing peaty burns

**Purple-loosestrife family,** *Lythraceae*
Water-purslane, *Lythrum portula*: edges of pools, rarely on peat

**Willowherb family,** *Onagraceae*
Broad-leaved Willowherb, *Epilobium montanum*: damp, shady places
Short-fruited Willowherb, *Epilobium obscurum*: variety of habitats
American Willowherb, *Epilobium ciliatum*: introduced; disturbed ground
Marsh Willowherb, *Epilobium palustre*: wet places
New Zealand Willowherb, *Epilobium brunnescens*: introduced; very successful coloniser of moist, open ground
Rose-bay Willowherb, *Chamerion angustifolium*: disturbed ground; invasive

**Holly family,** *Aquifoliaceae*
Holly, *Ilex aquifolium*: woodland and crags

**Spurge family,** *Euphorbiaceae*
Sun Spurge, *Euphorbia helioscopa*: cultivated and waste ground

**Flax family,** *Linaceae*
Fairy Flax, *Linum catharticum*: dry calcareous grassland
Allseed, *Radiola linoides*: damp, bare infertile ground

**Milkwort family,** *Polygalaceae*
Common Milkwort, *Polygala vulgaris:* calcareous grassland
Heath Milkwort, *Polygala serpyllifolia*: acid grassland and moorland; the commoner species locally, look for opposite leaves at the base of the stem

**Horse Chestnut family,** *Hippocastanaceae*
Horse Chestnut, *Aesculus hippocastanum*: planted

**Maple family,** *Aceraceae*
Sycamore, *Acer pseudoplatanus*: introduced; frequently planted; makes a fine tree, good for lichens

**Wood-sorrel family,** *Oxalidaceae*
Wood-sorrel, *Oxalis acetosella*: woodland and under bracken

**Crane's-bill family,** *Geraniaceae*
Wood Crane's-bill, *Geranium sylvaticum*: pre-1970 record
Cut-leaved Crane's-bill, *Geranium dissectum*: waste places
Small-flowered Crane's-bill, *Geranium pusillum*: sandy soils in open habitats; near its northern British limit
Dove's-foot Crane's-bill, *Geranium molle*: dry soils in open habitats
Herb-Robert, *Geranium robertianum*: shady, often rocky places
Stork's-bill, *Erodium cicutarium*: open grassland on Marian's Rock

**Ivy family,** *Araliaceae*
Ivy, *Hedera helix*: woodland and crags

**Carrot family,** *Apiaceae*
Marsh Pennywort, *Hydrocotyle vulgare*: wet places
Sanicle, *Sanicula europaea*: base-rich, moist soils in woodland
Cow Parsley, *Anthriscus sylvestris*: road verges and hay meadows
Sweet Cicely, *Myrrhis odorata*: introduced; road verges and river banks
Pignut, *Conopodium majus*: grassland, roadside verges and woodland

Ground Elder or Bishop's Weed, *Aegopodium podagraria*: garden and roadside weed, almost impossible to eradicate
Wild Angelica, *Angelica sylvestris*: damp woodland, grassland and roadsides; tall rounded, pinkish flower-heads in late summer
Hogweed, *Heracleum sphondylium*: roadsides and other tall herb communities; an important nectar source for insects
Upright Hedge-parsley, *Torilis japonica*: dry roadsides; here at its northern British limit
Wild Carrot, *Daucus carota*: calcareous, well-drained soils; uncommon in the east in northern Scotland

**Gentian family, *Gentianaceae***
Field Gentian, *Gentianella campestris*: open grassland; not uncommon in Corry Meadow; a northern species

**Periwinkle family, *Apocynaceae***
Lesser Periwinkle, *Vinca minor*: escape from cultivation; here at its northern British limit

**Bindweed family, *Convolvulaceae***
Hedge Bindweed, *Calystegia sepium*: introduced here, although native in much of Britain

**Bogbean family, *Menyanthaceae***
Bogbean, *Menyanthes trifoliata*: shallow pools, lochans, ditches and the old curling pond

**Borage family, *Boraginaceae***
Bugloss, *Anchusa arvensis*: arable fields and waste ground
Water Forget-me-not, *Myosotis scorpioides*: margins of streams and ponds, on calcareous soils
Creeping Forget-me-not, *Myosotis secunda*: margins of streams and pools, on acid soils
Tufted Forget-me-not, *Myosotis laxa*: wet, disturbed ground
Field Forget-me-not, *Myosotis arvensis*: dry, disturbed habitats
Changing Forget-me-not, *Myosotis discolor*: dry, open grassland and disturbed ground

**Deadnettle family, *Lamiaceae***
Hedge Woundwort, *Stachys sylvatica*: moist, shady places
Hybrid Woundwort, *Stachys x ambigua*: by streams, on road verges
Marsh Woundwort, *Stachys palustris*: marshes; the only woundwort whose leaves don't stink when bruised

Red Dead-nettle, *Lamium purpureum*: weed of cultivation
Large-flowered Hemp-nettle, *Galeopsis speciosa*: weed of cultivation
Common Hemp-nettle, *Galeopsis tetrahit*: weed of cultivation and disturbed habitats
Bifid Hemp-nettle, *Galeopsis bifida*: as *G. tetrahit*
Skullcap, *Scutellaria galericulata*: wetlands
Wood Sage, *Teucrium scorodonia*: crags, screes and mineral soils elsewhere
Bugle, *Ajuga reptans*: damp woodland and other shady places
Pyramidal Bugle, *Ajuga pyramidalis*: Nationally Scarce; a northern species; grassland, heaths and on crags, as near the Morvich Quarry; a pink-flowered form occurs locally
Ground Ivy, *Glechoma hederacea*: woodland and grassland; here close to its northern British limit; once used to flavour beer
Selfheal, *Prunella vulgaris*: moist grassland and open woodland
Wild Thyme, *Thymus polytrichus*: on rock outcrops and in open turf
Water Mint, *Mentha aquatica*: permanently wet places

**Mare's-tail family, *Hippuridaceae***
Mare's-tail, *Hippuris vulgaris*: at the edge of lochans

**Water-starwort family, *Callitrichaceae***
Common Water-starwort, *Callitriche stagnalis*: ephemeral pools and shallow permanent water
Intermediate Water-starwort, *Callitriche hamulata*: deep, still water and pools in fast-flowing rivers
Pedunculate Water-starwort, *Callitriche brutia*: ephemeral pools and shallow permanent water; thinly scattered in Britain

**Plantain family, *Plantaginaceae***
Sea Plantain, *Plantago maritima*: road verges and hill tops, some distance from the sea
Greater Plantain, *Plantago major*: road and tracksides, tolerant of trampling
Ribwort Plantain, *Plantago lanceolata*: grasslands and other places
Shoreweed, *Littorella uniflora*: submerged to 4m deep in lochans

**Ash family, *Oleaceae***
Ash, *Fraxinus excelsior*: moist, base-rich soils and cliffs; often planted
Wild Privet, *Ligustrum vulgare*: introduced in Scotland; here near its northern British limit

**Figwort family, *Scrophulariaceae***
Common Figwort, *Scrophularia nodosa*: damp, rich soils in woodland
Musk, *Mimulus moschatus*: introduced; waterside habitats
Monkeyflower, *Mimulus guttatus*: introduced; waterside habitats
Hybrid Monkeyflower, *Mimulus x robertsii*: partially-fertile hybrid; introduced; waterside habitats
Common Toadflax, *Linaria vulgaris*: waste places
Foxglove, *Digitalis purpurea*: disturbed ground in open situations and woodland
Thyme-leaved Speedwell, *Veronica serpyllifolia*: widespread
Heath Speedwell, *Veronica officinalis*: grassland and heathland
Germander Speedwell, *Veronica chamaedrys*: grassland
Marsh Speedwell, *Veronica scutellata*: wetlands
Brooklime, *Veronica beccabunga*: wetlands
Wall Speedwell, *Veronica arvensis*: open, dry habitats
Green Field-speedwell, *Veronica agrestis*: weed of cultivation
Slender Speedwell, *Veronica filiformis*: mown grasslands, including lawns
Common Cow-wheat, *Melampyrum pratense*: woods and heaths
Small Cow-wheat, *Melampyrum sylvaticum*: Nationally Scarce; 'Rogart Burn' (7302), MMcCW, 1951, here at its northern British limit
Eyebright, *Euphrasia officinalis*: grassland
Eyebright, *Euphrasia micrantha*: heaths and moorland
Yellow-rattle, *Rhinanthus minor*: nutrient-poor grasslands
Marsh Lousewort, *Pedicularis palustris*: lowland wetlands
Lousewort, *Pedicularis sylvatica*: heaths and moorland

**Bladderwort family, *Lentibulariaceae***
Butterwort, *Pinguicula vulgaris*: nutrient-poor bogs and base-rich flushes; insectivorous; starfish-like rosettes of pale-green leaves
Bladderwort, *Utricularia intermedia*: bog pools
Lesser Bladderwort, *Utricularia minor*: bog pools

**Bellflower family, *Campanulaceae***
Scottish Bluebell/Harebell, *Campanula rotundifolia*: dry grassland
Water Lobelia, *Lobelia dortmanna*: lochs with acid, stony substrates

**Bedstraw family, *Rubiaceae***
Field Madder, *Sherardia arvensis*: Marian's Rock (7401), P.Eaglesfield and MM, 1993; here near its northern British limit
Northern Bedstraw, *Galium boreale*: damp, base-rich, rocky places; a northern species

Sweet Woodruff, *Galium odoratum*: damp, base-rich soils in woodland; uncommon locally
Common Marsh-bedstraw, *Galium palustre*: wet meadows and marshes
Lady's Bedstraw, *Galium verum*: roadsides and rocky places
Hedge Bedstraw, *Galium mollugo*: base-rich grassland; here near its northern British limit
Heath Bedstraw, *Galium saxatile*: acid grassland and heaths
Sticky Willie/Cleavers, *Galium aparine*: fertile soils in a variety of habitats; can be a pestilential weed

**Honeysuckle family,** *Caprifoliaceae*
Elder, *Sambucus nigra*: usually found close to habitation; may be introduced here
Honeysuckle, *Lonicera periclymenum*: woodland and crags

**Moschatel family,** *Adoxaceae*
Moschatel, *Adoxa moschatellina*: also known as Town Hall Clock, due to its unusual inflorescence; sparingly, in hazel woodland; here near its northern British limit

**Valerian family,** *Valerianaceae*
Common Cornsalad, *Valerianella locusta*: dry, thin soils on crags, such as Marian's Rock; here near its northern British limit
Common Valerian, *Valeriana officinalis*: along the River Fleet

**Teasel family,** *Dipsacaceae*
Wild Teasel, *Dipsacus fullonum*: rough grassland and road verges; here at its northern British limit
Devil's-bit Scabious, *Succisa pratensis*: acid soils in a range of habitats

**Daisy family,** *Asteraceae*
Lesser Burdock, *Arctium minus*: rough grassland and road verges
Spear Thistle, *Cirsium vulgare*: rough grassland and disturbed habitats
Melancholy Thistle, *Cirsium heterophyllum*: thornless thistle; along the River Fleet; a northern species
Marsh Thistle, *Cirsium palustre*: wet places
Creeping Thistle, *Cirsium arvense*: disturbed ground
Cornflower, *Centaurea cyanus*: weed of cereal fields, recorded since 1970; here near its northern British limit
Knapweed, *Centaurea nigra*: meadows, road verges and waste ground
Nipplewort, *Lapsana communis*: shady, disturbed ground

Cat's-ear, *Hypochaeris radicata*: dry grassland
Autumn Hawkbit, *Leontodon autumnalis*: dry grassland
Perennial Sow-thistle, *Sonchus arvensis*: road verges, arable and waste places; showy, shaggy flowers
Smooth Sow-thistle, *Sonchus oleraceus*: waste places
Prickly Sow-thistle, *Sonchus asper*: dry, disturbed soils
Dandelion, *Taraxacum officinale*: disturbed and marshy ground
Marsh Hawk's-beard, *Crepis paludosa*: rocky, wooded streamsides and gullies; a northern species
Smooth Hawk's-beard, *Crepis capillaris*: early coloniser of open, dry habitats
Mouse-ear Hawkweed, *Pilosella officinarum*: dry, short grassland
Fox-and-cubs, *Pilosella aurantiaca*: escape from cultivation; followed railway line north?
Hawkweed, *Hieracium reticulatum*: rocky places; Scottish species
Hawkweed, *Hieracium strictiforme*: rocky places; northern species
Hawkweed, *Hieracium saxorum*: rocky places; British endemic
Common Cudweed: *Filago vulgaris*: dry, open habitats; a southern species, here near its northern British limit
Small Cudweed, *Filago minima*: dry, open habitats; here near its northern British limit
Mountain Everlasting, *Antennaria dioica*: thin soils in heath and on rocks; locally common in upland areas
Heath Cudweed, *Gnaphalium sylvaticum*: open, dry, acid soils
Marsh Cudweed, *Gnaphalium uliginosum*: muddy ground in fields and along roadsides; tolerant of trampling
Golden-rod, *Solidago virgaurea*: rock outcrops and hill grasslands
Daisy, *Bellis perennis*: damp, heavily grazed grassland
Tansy, *Tanacetum vulgare*: waste ground; escape from cultivation?
Sneezewort, *Achillea ptarmica*: damp places
Yarrow, *Achillea millefolium*: dry grassland
Corn Marigold, *Chrysanthemum segetum*: weed of cereal fields; pre-1970 record
Ox-eye Daisy, *Leucanthemum vulgare*: dry grassland, as on roadsides
Pineappleweed, *Matricaria discoidea*: introduced; roadsides
Sea Mayweed, *Tripleurospermum maritimum*: roadsides
Scentless Mayweed, *Tripleurospermum inodorum*: weed of farmland and waste ground
Ragwort, *Senecio jacobaea*: overgrazed pasture
Marsh Ragwort, *Senecio aquaticus*: wet ground
Groundsel, *Senecio vulgaris*: cultivated and waste ground

Heath Groundsel, *Senecio sylvaticus*: sandy soils in open habitats
Sticky Groundsel, *Senecio viscosus*: disturbed ground
Coltsfoot, *Tussilago farfara*: damp disturbed ground and river shingle; flowering early before the leaves appear

**Arrowgrass family, *Juncaginaceae***
Marsh Arrowgrass, *Triglochin palustre*: marshy places

**Pondweed family, *Potamogetonaceae***
Broad-leaved Pondweed, *Potamogeton natans*: still or slow-moving waters
Bog Pondweed, *Potamogeton polygonifolius*: burns, bog pools and *Sphagnum* lawns
Various-leaved Pondweed, *Potamogeton gramineus*: shallow water; scattered and northern in Britain
Red Pondweed, *Potamogeton alpinus*: lochs and burns, where silt accumulates

**Duckweed family, *Lemnaceae***
Common Duckweed, *Lemna minor*: still or slow-moving waters; rare in northern Scotland away from the east coast

**Rush family, *Juncaceae***
Heath Rush, *Juncus squarrosus*: wet, peaty heaths and moorland
Slender Rush, *Juncus tenuis*: introduced; rare in northern Scotland
Toad Rush, *Juncus bufonius*: winter-wet, open habitats
Jointed Rush, *Juncus articulatus*: wet places
Sharp-flowered Rush, *Juncus acutiflorus*: acid wet habitats
Bulbous Rush, *Juncus bulbosus*: in or by water, or on wet peat
Soft Rush, *Juncus effusus*: wet ground and neglected grassland
Compact Rush, *Juncus conglomeratus*: acid, sandy or peaty damp ground
Hairy Wood-rush, *Luzula pilosa*: woodlands and heathlands
Great Wood-rush, *Luzula sylvatica*: gorges and woodlands
Field Wood-rush, *Luzula campestris*: short, infertile grassland
Heath Wood-rush, *Luzula multiflora*: acid grassland

**Sedge family, *Cyperaceae***
Common Cottongrass, *Eriophorum angustifolium*: wet heath and bog pools
Hare's-tail Cottongrass, *Eriophorum vaginatum*: mires and blanket bogs
Deergrass, *Trichophorum caespitosum*: wet moorland

Common Spike-rush, *Eleocharis palustris*: margins of pools
Few-flowered Spike-rush, *Eleocharis quinqueflora*: base-rich flushes
Common Club-rush, *Schoenoplectus lacustris*: standing or flowing water
Bristle Club-rush, *Isolepis setacea*: tracks and loch shores
Floating Club-rush, *Eleogiton fluitans*: bog pools and lochans
Black Bog-rush, *Schoenus nigricans*: base-rich fens, flushes and marshes; predominantly north-western in Britain
Lesser Tussock-sedge, *Carex diandra*: wet, peaty places
Oval Sedge, *Carex ovalis*: acid grassland and heath
Star Sedge, *Carex echinata*: waterlogged sites
Dioecious Sedge, *Carex dioica*: wet, base-rich habitats
White Sedge, *Carex curta*: mires and bogs
Slender Sedge, *Carex lasiocarpa*: reed-swamps and mires
Bottle Sedge, *Carex rostrata*: pools in burns and lochans
Glaucous Sedge, *Carex flacca*: neutral grassland and flushes
Carnation Sedge, *Carex panicea*: widespread in damp habitats
Green-ribbed Sedge, *Carex binervis*: acid grassland and heath
Tawny Sedge, *Carex hostiana*: base-rich flushes
Yellow Sedge, *Carex viridula* ssp. *brachyrrhyncha*: base-rich flushes
Yellow Sedge, *Carex viridula* ssp. *oedocarpa*: wet, open habitats
Pale Sedge, *Carex pallescens*: damp woods and grassland
Pill Sedge, *Carex pilulifera*: acidic, dry soils
Bog-sedge, *Carex limosa*: peaty pools
Common Sedge, *Carex nigra*: wide variety of wet habitats
Few-flowered Sedge, *Carex pauciflora*: bogs; northern species
Flea Sedge, *Carex pulicaris*: damp neutral soils

**Grass family, *Poaceae***
Mat Grass, *Nardus stricta*: winter-wet, acid, infertile soils
Giant Fescue, *Festuca gigantea*: damp woodland on neutral soils; here close to its northern British limit
Red Fescue, *Festuca rubra*: all kinds of grassland
Sheep's Fescue, *Festuca ovina*: infertile, well-drained grassland
Fine-leaved Sheep's-fescue, *Festuca filiformis*: acid, sandy, well-drained soils on heaths and moors
Perennial Rye-grass, *Lolium perenne*: pastures and roadsides
Italian Rye-grass, *Lolium multiflorum*: agricultural grasslands
Squirreltail Fescue, *Vulpia bromoides*: open habitats
Rat's-tail Fescue, *Vulpia myuros*: waste ground; very rare in northern Scotland; one record, 1961, MMcCW
Crested Dog's-tail, *Cynosurus cristatus*: agricultural and other grasslands

Reflexed Saltmarsh-grass, *Puccinellia distans*: has recently colonised salt-treated roads; otherwise a coastal species
Annual Meadow-grass, *Poa annua*: man-made habitats; tolerant of trampling
Rough Meadow-grass, *Poa trivialis*: grasslands
Spreading Meadow-grass, *Poa humilis*: roadsides and other habitats
Smooth Meadow-grass, *Poa pratensis*: well-drained, neutral grasslands
Wood Meadow-grass, *Poa nemoralis*: shady places, uncommon in north
Cock's-foot, *Dactylis glomerata*: wide range of grasslands
Floating Sweet-grass, *Glyceria fluitans*: floating rafts in shallow water
Downy Oat-grass, *Helictotrichon pubescens*: base-rich grasslands
False Oat-grass, *Arrhenatherum elatius*: roadsides and rough grassland
Wild Oat, *Avena fatua*: weed of cereals; pre-1970 record
Crested Hair-grass, *Koeleria macrantha*: base-rich grasslands
Tufted Hair-grass, *Deschampsia caespitosa*: poorly-drained soils; tussock-forming
Wavy Hair-grass, *Deschampsia flexuosa*: heaths and moorland
Yorkshire Fog, *Holcus lanatus*: damp grasslands
Creeping Soft-grass, *Holcus mollis*: woodlands
Silver Hair-grass, *Aira caryophyllea*: sandy and rocky places; a tiny, dainty grass
Early Hair-grass, *Aira praecox*: rock outcrops
Sweet Vernal Grass, *Anthoxanthum odoratum*: acid grasslands
Reed Canary-grass, *Phalaris arundinacea*: wet places where the water table fluctuates
Common Bent, *Agrostis capillaris*: acid grasslands
Creeping Bent, *Agrostis stolonifera*: damp grasslands and water margins
Velvet Bent, *Agrostis canina*: peaty soils, mires and wet heaths
Brown Bent, *Agrostis vinealis*: dry, acid, sandy or peaty soils
Meadow Foxtail, *Alopecurus pratensis*: moist, fertile grasslands
Marsh Foxtail, *Alopecurus geniculatus*: fertile sites flooded in winter
Timothy, *Phleum pratense*: damp, fertile soils
Soft-brome, *Bromus hordeaceus*: disturbed, fertile, neutral soils
Slender Soft-brome, *Bromus lepidus*: improved grasslands
False Brome, *Brachypodium sylvaticum*: well-drained woodland
Common Couch, *Elytrigia repens*: fertile, disturbed ground
Heath-grass, *Danthonia decumbens*: heath grassland and moorland
Purple Moor-grass, *Molinia caerulea*: wide range of damp habitats; very common on hill ground
Common Reed, *Phragmites australis*: forms large stands in shallow water and at the edge of lochans

**Bur-reed family,** *Sparganiaceae*
Branched Bur-reed, *Sparganium erectum*: nutrient-rich shallow water
Unbranched Bur-reed, *Sparganium emersum*: deeper water

**Lily family,** *Liliaceae*
Bog Asphodel, *Narthecium ossifragum*: wet heaths and mires
Bluebell, *Hyacinthoides non-scripta*: woodland, under bracken and on cliffs
Ramsons/Wild Garlic, *Allium ursinum*: moist woodlands, riversides, rock crevices

**Iris family,** *Iridaceae*
Yellow Iris, *Iris pseudacorus*: wet meadows, woods and loch margins

**Orchid family,** *Orchidaceae*
Bird's-nest Orchid, *Neottia nidus-avis*: woodland near Torbreck (7003), M.J.Marshall, 1978; here near its northern British limit on the east coast
Lesser Twayblade, *Listera cordata*: in *Sphagnum* under heather; a northern species
Lesser Butterfly Orchid, *Platanthera bifolia*: range of damp habitats
Fragrant Orchid, *Gymnadenia conopsea*: calcareous grassland, heath and fens; distinctive colour and scent
Common Spotted Orchid, *Dactylorhiza fuchsii*: base-rich soils in a range of habitats
Heath Spotted Orchid, *Dactylorhiza maculata* ssp. *ericetorum*: grassland, moors, heaths, flushes and bogs; our commonest orchid
Northern Marsh Orchid, *Dactylorhiza purpurella*: damp, base-rich soils.

# Bryophytes
*Compiled by Ian Evans*

Bryophytes include mosses, liverworts and hornworts, 'small, green photosynthetic plants that do not produce flowers, seeds or fruits' (Porley and Hodgetts 2005). Some have distinct stems and leaves, others (the thalloid liverworts) flat plates of green tissue. Most disperse themselves by means of spores. Because they are relatively small, they do not generally compete well with the larger flowering plants and ferns. However, they are pioneer colonists (with algae and lichens) of 'bare' habitats such as soil, rock, trees and some man-made structures. They may also be the dominant life form in bogs, mires and some montane communities, and are an important component of the vegetation of other mature habitats such as grassland, heathland, the edges of lochs and burns, woodland, rock and scree, extending from sea-level to the tops of the highest hills. They are also very beautiful when examined with a hand lens, or, better still, a microscope.

The bryophyte communities of the wetter western seaboard of Highland Scotland are of international importance and parts of this area have been well surveyed; see, for example, Gordon Rothero's account of the bryophytes of Assynt in West Sutherland (Evans, Evans and Rothero 2002). However, until very recently, East Sutherland had not been so well explored, and the area around Rogart was a good example. The 10km square NC70, which stretches from Rogart north to Strath Brora, mustered, up to 2006, just 53 species of mosses, including only three species of *Sphagnum*, and 20 species of liverworts, less than a half of the number that might be expected.

Members of the British Bryological Society held a field meeting at Golspie in early July 2006 to help remedy this situation, visiting a number of localities in East Sutherland. On $7^{th}$ July, five of them, John Blackburn, Richard Fisk, Mary Ghullam, Mark Pool and Jo Wilbraham, had a look at the bryophytes of four areas in the vicinity of Rogart and the tables which follow (pages 104-106) detail what they found.

The areas and habitats recorded were:
1. Roadsides and walls in Rogart village, from the crossroads down to the station, and the adjacent burn course (7201).
2. Walls, banks and trees on the road east of the crossroads, towards Remusaig and the adjacent woodland along the Garbh-allt or Corry Burn (7202/7302).

3. Woodland, heathland, rock outcrops, streamsides and flushes to the west of Millnafua Bridge (7202/7302).
4. Hazel and birch woodland in the gorge of the upper part of the Garbh-allt at Little Rogart (7304).

The group makes no claim that the lists were exhaustive; one visit cannot hope to encompass all the available habitats in an area as varied in its topography as Rogart, let alone the whole of a 10km square. However, the total number of species recorded was a very respectable 159, comprising 128 mosses and 31 liverworts.

A few species previously listed from NC70 were not re-found on this occasion, although it is not known, at present, exactly where they were originally recorded; some of the records were made prior to 1950. They comprise the mosses *Anomobryum julaceum*, *Calliergon cordifolium*, *Calliergon stramineum*, *Ctenidium molluscum*, *Dicranella crispa* (uncommon), *Dicranum fuscescens*, *Fontinalis antipyretica*, *Grimmia decipiens*, *Hookeria lucens* and the liverworts *Cephalozia leucantha* (pre-1950 record, uncommon), *Cephaloziella rubella*, *Jungermannia obovata*, *Marchantia polymorpha*, *Scapania subalpina*. With others mentioned below, they bring the total for NC70 up to 179 species.

Gordon Rothero, who has extensive experience of Highland bryophytes, comments that the occurrence, in the 2006 lists, of species such as *Sphagnum warnstorfii* and *Drepanocladus cossonii* indicate some base richness in flushes, but that some of the commoner species of base-rich rocks are missing [although, one such, *Ctenidium molluscum*, had been recorded previously]. He goes on to say that the records of *Orthotrichum pulchellum* and, to a lesser extent, the other species of *Orthotrichum*, convey 'an eastern tinge to the list' as does the absence of many oceanic species so familiar in Assynt, such as *Scapania gracilis*.

While a good start has now been made on recording the bryophytes of the Rogart area (and the rest of NC70), there is plenty of scope for further exploration. This is born out by a remarkable discovery made just to the east of the village, by two experienced members of the B.B.S. on their way home from the field meeting.

Sam Bosanquet says that he had had his eye on the granite outcrops that tower above the road since acquiring the O.S. maps for the area. He and Chris Preston decided to take 'a quick look' on their way south. He had 'been hoping for *Grimmia* and *Schistidium* [cushion-forming mosses, usually on rocks], but wasn't expecting such riches. There were a number of patches of a large but totally unfamiliar *Grimmia* ... covering several square metres'; this turned out to be *G. elatior*, which was first discovered in Glen Clova in 1868, and had not been seen there, or anywhere else in the British Isles, since 1871. In its vicinity was 'abundant *Grimmia decipiens* and various other species, including *G. funalis*, *G.trichophylla* and the '*retracta*' form of *G. lisae*'. He collected two *Schistidium* species, 'which turned out to be *S. pruinosum* and *S. dupretii*, both of which grow at a single site on the Old Red Sandstone further north in East Sutherland'. The last two species are both scarce in Scotland, and almost absent from England and Wales. The liverwort *Porella obtusata*, new to East Sutherland, was also present, and another member of the B.B.S., Mark Lawley, later added yet another species of *Grimmia*, *G. ovalis*, on his way south. [It should be said that *Grimmia* and *Schistidium* are amongst the more challenging groups of mosses, even for experienced bryologists.]

**References**

Evans, P.A., Evans, I.M. and Rothero, G.P., 2002. *Flora of Assynt*. Privately published. [Contains a substantial section, by Gordon Rothero, on the bryophytes of this area of West Sutherland, the first study at this level of detail of any Highland parish].

Porley, R. and Hodgetts, N., 2005. *Mosses and Liverworts*. London: Collins New Naturalist. [An authoritative and very readable account of the natural history of the group, beautifully illustrated.]

**Overleaf. Bryophyte species/habitat lists: for location of habitats see pages 101-102.**

| Moss species | 1 | 2 | 3 | 4 | Moss species | 1 | 2 | 3 | 4 |
|---|---|---|---|---|---|---|---|---|---|
| Amphidium mougeotii |  |  | * |  | Dicranella staphylina | * |  |  |  |
| Antitrichia curtipendula |  |  | * |  | Dicranella varia |  |  |  | * |
| Atrichum undulatum | * |  | * | * | Dicranoweisia cirrata |  | * |  |  |
| Aulacomnium androgynum | * |  |  |  | Dicranum majus |  |  |  | * |
| Aulacomnium palustre |  | * |  |  | Dicranum scoparium | * | * | * | * |
| Barbula convoluta | * | * | * | * | Didymodon insulanus | * | * |  |  |
| Barbula unguiculata | * | * |  |  | Didymodon rigidulus | * | * |  |  |
| Bartramia pomiformis |  |  |  | * | Ditrichum cylindricum |  | * | * |  |
| Blindia acuta |  |  | * |  | Drepanocladus cossonii |  |  | * |  |
| Brachythecium plumosum | * | * | * |  | Drepanocladus revolvens |  |  | * |  |
| Brachythecium rivulare | * | * |  | * | Encalypta streptocarpa | * |  |  |  |
| Brachythecium rutabulum |  | * | * | * | Entosthodon obtusus |  |  | * |  |
| Breutelia chrysocoma |  |  | * |  | Eurhynchium hians |  |  |  | * |
| Bryum alpinum |  |  | * |  | Eurhynchium praelongum |  | * | * | * |
| Bryum argenteum |  | * |  |  | Eurhynchium striatum | * |  |  | * |
| Bryum bicolor | * | * |  |  | Fissidens adianthoides |  | * | * |  |
| Bryum capillare | * | * | * |  | Fissidens taxifolius |  |  | * |  |
| Bryum pallens |  |  | * |  | Funaria hygrometrica |  | * |  |  |
| Bryum pallescens |  | * |  |  | Grimmia curvata |  |  | * |  |
| Bryum pseudotriquetrum | * |  | * |  | Grimmia hartmanii | * |  |  |  |
| Bryum rubens |  | * |  |  | Grimmia pulvinata | * | * |  |  |
| Bryum ruderale |  | * |  |  | Grimmia trichophylla |  |  | * | * |
| Calliergonella cuspidata | * | * | * |  | Gymnostomum aeruginosum |  |  |  | * |
| Campylium stellatum var. stellatum |  | * |  |  | Hedwigia ciliata | * |  | * |  |
| Campylopus atrovirens |  |  | * |  | Hedwigia stellata | * | * |  |  |
| Campylopus introflexus |  |  | * |  | Homalothecium sericeum | * | * |  |  |
| Ceratodon purpureus | * | * | * |  | Hylocomium brevirostre |  |  |  | * |
| Cirriphyllum piliferum |  | * |  | * | Hylocomium splendens | * | * | * | * |
| Climacium dendroides | * |  | * |  | Hypnum andoi | * |  | * | * |
| Dichodontium pellucidum |  |  |  | * | Hypnum cupressiforme |  | * | * | * |
| Dicranella heteromalla |  |  |  | * | Hypnum jutlandicum |  | * |  |  |
| Dicranella palustris |  |  | * |  | Hypnum resupinatum | * | * |  |  |
| Dicranella rufescens |  |  |  | * | Isothecium alopecuroides | * | * | * | * |

| Moss species | Habitats | | | | Moss species | Habitats | | | |
|---|---|---|---|---|---|---|---|---|---|
| | 1 | 2 | 3 | 4 | | 1 | 2 | 3 | 4 |
| Isothecium myosuroides | * | | * | * | Racomitrium ericoides | * | | | |
| Isothecium myosuroides var. myosuroides | | * | | | Racomitrium fasciculare | * | * | * | |
| | | | | | Racomitrium heterostichum | * | * | * | |
| Leucobryum glaucum | | | * | | Racomitrium lanuginosum | * | | * | |
| Mnium hornum | * | * | * | * | Rhizomnium punctatum | | | | * |
| Neckera complanata | | | * | * | Rhynchostegium riparioides | * | * | | * |
| Orthotrichum affine | | * | * | | Rhytidiadelphus loreus | | | * | * |
| Orthotrichum anomalum | | * | | | Rhytidiadelphus squarrosus | * | * | * | |
| Orthotrichum pulchellum | * | * | * | * | Rhytidiadelphus triquetrus | * | | * | * |
| Orthotrichum stramineum | | * | | | Schistidium apocarpum | | * | | |
| Orthotrichum striatum | * | * | * | * | Scleropodium purum | * | | * | * |
| Palustriella commutata var. commutata | | | * | | Scorpidium scorpioides | | * | | |
| | | | | | Sphagnum capillifolium subsp. rubellum | | | * | |
| Philonotis fontana | * | * | | | | | | | |
| Plagiomnium elatum | | | * | | Sphagnum contortum | | | * | |
| Plagiomnium undulatum | | * | * | * | Sphagnum denticulatum | | | * | |
| Plagiothecium undulatum | | | * | * | Sphagnum palustre | | | * | |
| Pleurozium schreberi | * | | * | * | Sphagnum papillosum | | | * | |
| Pogonatum aloides | | | * | | Sphagnum subnitens | | | * | |
| Pogonatum urnigerum | * | * | * | * | Sphagnum tenellum | | | * | |
| Pohlia annotina | | | | * | Sphagnum warnstorfii | | | * | |
| Pohlia drummondii | | | | * | Syntrichia ruralis | * | * | | |
| Pohlia melanodon | | * | | | Thamnobryum alopecurum | * | * | | * |
| Pohlia nutans | | | | * | Thuidium tamariscinum | * | * | * | * |
| Pohlia wahlenbergii | | | * | * | Tortella tortuosa | * | | | |
| Polytrichum formosum | * | | * | * | Tortula muralis var. muralis | * | * | | |
| Polytrichum juniperinum | * | | * | | Tortula truncata | | * | | |
| Polytrichum piliferum | * | | * | | Trichostomum brachydontium | | | * | |
| Pseudocrossidium hornschuchianum | | * | | | Ulota bruchii | * | * | * | |
| Pseudocrossidium revolutum | | * | | | Ulota crispa | | | * | * |
| | | | | | Ulota drummondii | * | * | * | * |
| Racomitrium aciculare | | * | * | * | Ulota phyllantha | * | * | * | * |
| Racomitrium affine | | | * | | | | | | |

| Liverwort species | Habitats | | | | Liverwort species | Habitats | | | |
|---|---|---|---|---|---|---|---|---|---|
| | 1 | 2 | 3 | 4 | | 1 | 2 | 3 | 4 |
| Aneura pinguis | | | * | | Metzgeria furcata | * | * | * | * |
| Barbilophozia hatcheri | | | * | | Mylia anomala | | | * | |
| Calypogeia fissa | | | * | | Nardia scalaris | * | | * | * |
| Calypogeia muelleriana | | | | * | Pellia endiviifolia | | | * | |
| Cephalozia bicuspidata | | | * | * | Pellia epiphylla | | | * | |
| Chiloscyphus pallescens | | | * | | Pellia neesiana | | | * | |
| Chiloscyphus polyanthos | * | * | | * | Plagiochila asplenioides | | | * | * |
| Conocephalum conicum | * | * | | * | Plagiochila porelloides | * | * | | |
| Diplophyllum albicans | | | | * | Porella cordaeana | | | * | |
| Fossombronia wondraczekii | | | * | | Radula complanata | * | * | * | * |
| Frullania dilatata | | * | | | Riccardia multifida | | | * | |
| Frullania tamarisci | * | * | * | * | Riccia sorocarpa | | | * | * |
| Harpanthus flotovianus | | | * | | Scapania compacta | * | | | |
| Lejeunea cavifolia | | | | * | Scapania umbrosa | | | | * |
| Lepidozia reptans | | | | * | Scapania undulata | * | | | * |
| Lophocolea bidentata | * | | * | * | | | | | |

# Fungi
*John Blunt*

**Introduction**

Fungi were once regarded as 'lower plants', but are now recognised as a separate Kingdom of living organisms. They play an extremely important role in the living world as a whole, whether in the symbiotic partnerships with most plants known as mycorrhizae, as parasites or decomposers. Lichens are another example of a symbiotic partnership, usually between ascomycete fungi and cyanobacteria or green algae. A few fungi are parasitic on animals.

The most obvious fungi are the fruiting bodies of 'mushrooms, toadstools' and their allies, often large and colourful, which are produced by an extensive network of fungal threads ramifying through a substrate such as soil or decaying wood. These fungi, belonging to a group called the Basidiomycetes, represent only about a quarter of the 12,000 or so species that have so far been described from the British Isles, but they are a prominent component of the numerous coloured guides to the group.

Putting a definite name to any but a handful of fungi is not an easy matter. Material must be carefully collected, with habitat details, and collecting can involve a lot of time on one's hands and knees! Back at home, specimens must be laid out so that spore prints can be taken and parts of the fruiting bodies and the spores must then be examined under a high power microscope, at magnifications up to 1000 times, with chemical tests as appropriate. To add to the difficulties, many groups of species are only described in detail in the specialist journals. It can take a couple of hours to name a specimen from one of the 'awkward' groups. Not surprisingly, mycologists tend to specialise in, for instance, agarics, polypores, cup and disc fungi or lichens.

Given the challenges, it is perhaps no wonder that it takes a minimum of five years to make a representative list from an area as small as a 1km square, with 10-12 visits a year, spread through the seasons. However such an area, with a good mixture of habitats, may yield 1000 species, compared with perhaps 200 of higher plants.

The lists that follow are the product of visits, in the company of Ian Evans, to a range of habitats spread over ten 1km squares in the Rogart area, on 16.9.06, 20.10.06 and 29.10.06. These three visits yielded 182 records of some 140 species, which is a respectable total for one amateur mycologist with limited expertise and indicates that **the area is very rich indeed for fungi**. For comparison, a 10 km. square on the West Coast, which the author has worked over the last 25 years, has so far yielded just 1550 records of some 480 species.

At least 10% of the fungi seen and collected could not be named, for a variety of reasons. Certain groups were beyond the competence of the author (and many other mycologists!); they included the genera *Cortinarius* and *Inocybe*, and nearly all the resupinate basidiomycetes; specimens of others did not yield spores, decomposed or were consumed by invertebrates.

Most of the records are of **basidiomycetes**, the group that includes the gill and bracket fungi, whose fruiting bodies are particularly conspicuous in the autumn. There are a fair number of **ascomycetes**, cup fungi and their allies, in which the author has a special interest, and a few from other groups, the phycomycetes and coelomycetes. The species listed represent a fair cross-section of ecotypes. Examples of the mycorrhizal species include many members of the genera *Boletus*, *Lactarius*, *Russula* and *Tricholoma*. Parasites include the two conspicuous bracket fungi on birch, *Fomes fomentarius* and *Piptoporus betulinus*, and ascomycetes in the genera *Rhytisma* and *Taphrina*, which occur on a variety of trees. Decomposers are represented by most of the ascomycetes recorded and basidiomycete genera such as *Collybia*, *Exidia* and *Chondrostereum*.

The lists that follow indicate that the different habitats sampled do have distinctive fungal assemblages, although some common species, such as the mycorrhizal False Chanterelle *Hygrophoropsis aurantiaca* and the wood-decomposing ascomycete *Bisporella citrina*, were found in several of them.

The lists for each locality are divided into basidiomycetes (with names based on the recent checklist by Legon and Henrici, 2005), ascomycetes (with names based on the on-line B.M.S. checklist) and other groups. English names have been added for most of the basidiomycetes and a few of the ascomycetes. They are taken from the new edition of Roger Phillips' *Mushrooms* (2006), which contains illustrations of many of those listed.

The main localities are arranged from south to north, with a composite list at the end for records from other localities. An electronic version of the list is available from the author. It is hoped to add to the lists for the area in years to come and any records from interested parties will be gratefully received by the author at: Burnside, Nedd, Drumbeg, by Lairg, Sutherland, IV27 4NN.

## Lists of species recorded

### 1. Conifer plantation south of Davochbeg, NC7201, visited 16.10.06

**Basidiomycetes**
Buttercap, *Collybia butyracea*: conifer litter; litter decomposer
False Chanterelle, *Hygrophoropsis aurantiaca*: under conifers: litter decomposer
Shaggy Parasol, *Macrolepiota rhacodes*: under conifers; litter decomposer
Bonnet, *Mycena cinerella*: pine litter; litter decomposer
Bleeding Bonnet, *Mycena sanguinolenta*: conifer debris; litter decomposer
Flaming Scalycap, *Pholiota flammans*: rotten wood, probably pine; wood decomposer
*Ramaria flaccida*: amongst conifer litter; mycorrhizal or saprophytic
Larch Bolete, *Suillus grevillei*: under larch; mycorrhizal

**Ascomycetes**
*Bulgariella pulla*: conifer wood, probably pine; wood decomposer
*Gorgoniceps aridula*: pine bark; wood decomposer
*Lophodermium pini-excelsae*: pine leaves; host-specific leaf decomposer

### 2. Davochbeg garden, NC7201, visited 26.9.06, 16.10.06 and 29.10.06. Lawns and flower/shrub borders surrounded by mature trees of beech, common lime, sycamore, wych elm etc.

**Basidiomycetes**
Field Mushroom, *Agaricus campestris*: 26.9
Scaly Wood Mushroom, *Agaricus langei*: 16.10
Dark Fieldcap, *Agrocybe erebia*: 16.10
Red Cracking Bolete, *Boletus chrysenteron*: fine show of fruiting bodies on 26.9
Deceiving Bovist/Puffball, *Bovista aestivalis*: 24.10 [only two previous records from Highland]
Inkcap, *Coprinus leiocephalus*: 29.10
Earthy Powdercap, *Cystoderma amianthinum*: 26.9

*Exidia thuretina*: on unidentified dead wood, 29.10
Collared Parachute, *Marasmius rotula*: on woody material, under beech, 26.9
Nitrous Bonnet, *Mycena leptocephala*: 26.9
Grooved Bonnet, *Mycena polygramma*: on dead wood, 29.10
Golden Bootleg, *Phaeolepiota aurea*: 16.10
Meadow Puffball, *Vascellum pratense*: 26.9

**Ascomycetes**
*Hymenoscyphus rokebyensis*: on beechmast, 26.9
Coral Spot, *Nectria cinnabarina*: on dead beech branch, 29.10
Tan Ear, *Otidea alutacea*: 29.10
*Phaeohelotium trabellinum*: on beechmast, 29.10
Sycamore Tarspot, *Rhytisma acerinum*: on sycamore leaves, 29.10
*Rosellinia aquila*: on bark of beech, 29.10

**3. Corry Meadow, Ard a'Chlachain, old curling pond and associated areas, NC7201/7301, visited 26.9.06 and 29.10.06. A rich area with a great variety of habitats including acid grassland/dry heath, birch woodland, marshy ground and wet woodland associated with the old curling pond, and some large oaks.**

**Basidiomycetes**
Yellow Club, *Clavulinopsis helvola*: acid grassland/heath, 7301, 26.9
Fragrant Funnel, *Clitocybe fragrans*: in litter under birch, 7301, 29.10
Clouded Funnel, *Clitocybe nebularis*: under birch, 7201, 29.10
Russet Toughshank, *Collybia dryophila*: acid grassland/heath, 7201, 26.9
Flat Oysterling, *Crepidotus applanatus*: on large log, 7301, 26.9
Common Jellyspot, *Dacrymyces stillatus*: on oak wood, 7301, 26.9
Pinkgill, *Entoloma prunaloides*: acid grassland/heath, 7301, 26.9
Witches' Butter, *Exidia glandulosa*: on oak wood, 7301, 29.10
*Exidia repanda*: on wood of birch, 7301, 29.10
Tinder Bracket or Hoof Fungus, *Fomes fomentarius*: on birch, 7301, 29.10
*Galerina atkinsoniana*: acid grassland/heath, 7201, 26.9
Waxcap, *Hygrocybe colemaniana*: acid grassland/heath, 7301, 29.10
Blackening Waxcap, *Hygrocybe conica*: acid grassland/heath, 7301, 26.9
Waxcap, *Hygrocybe mucronella*: acid grassland/heath, 7301, 26.9
Meadow Waxcap, *Hygrocybe pratensis*: acid grassland/heath, 7301, 29.10
Crimson Waxcap, *Hygrocybe punicea*: acid grassland/heath, 7301, 26.9
Waxcap, *Hygrocybe reidii*: acid grassland/heath, 7301, 26.9

Amethyst Deceiver, *Laccaria amethysina*: under birch/willow, 7301, 29.10
The Deceiver, *Laccaria laccata*: under birch, 7301, 26.9
Ugly Milkcap, *Lactarius turpis*: under birch, 7301, 26.9
Brown Birch Bolete, *Leccinum scabrum*: under birch, 7301, 26.9
Orange Birch Bolete, *Leccinum versipelle*: under birch, 7301, 29.10
*Melampsora populnea*: rust, on aspen leaves, 7201, 29.10
*Merismodes anomalus*: dead wood, 7201, 29.10
Yellowleg Bonnet, *Mycena epipterygia*: very rotten wood, 7301, 29.10
Grooved Bonnet, *Mycena polygramma*: dead wood, 7301, 29.10
Willow Bracket, *Phellinus igniarius*: on willow wood, 7301, 29.10
Bay Polypore, *Polyporus durus*: on willow wood, 7301, 29.10
*Psilocybe merdicola*: on horse dung, 7301, 29.10
Liberty Cap, *Psilocybe semilanceata*: in grass, 7201, 26.9; acid grassland/heath, 7301, 26.9
*Puccinia violae*: rust, on leaves of common dog-violet, 7301, 29.10
Charcoal Burner, *Russula cyanoxantha*: under birch, 7301, 26.9
Bleached Brittlegill, *Russula exalbicans*: under birch, 7301, 26.9
Hairy Curtain Crust, *Stereum hirsutum*: on oak wood, 7301, 29.10
Dung Roundhead, *Stropharia semiglobata*: acid grassland/heath, 7301, 26.9
*Tricholoma stiparophyllum*: under birch, 7301, 26.9

**Ascomycetes**
*Bisporella citrina*: dead wood, 7301, 29.10
Green Elfcup, *Chlorociboria aeruginascens*: dead wood, 7301, 29.10
*Diatrype bullata*: on willow wood, 7301, 29.10
*Hyaloscypha hyalina*: dead wood, 7301, 29.10
*Hypoxylon multiforme*: on birch wood, 7301, 29.10
*Lachnum brevipilosum*: dead wood, 7301, 29.10
*Lachnum pygmaeum*: on gorse wood, 7301, 29.10
*Mollisia cinerea*: dead wood, 7201, 29.10
*Nemania serpens*: dead wood, 7301, 29.10
*Phaeohelotium italicum*: dead wood, 7302, 29.10
*Taphrina tormentillae*: on tormentil, acid grassland, 7201, 26.9
Candlesnuff Fungus, *Xylaria hypoxylon*: dead wood, 7301, 29.10

**4. Birch and juniper woodland above back road at Pitfure East, NC7103, visited 16.10.06**

**Basidiomycetes**
Fly Agaric, *Amanita muscaria*: under birch; mycorrhizal

Panthercap, *Amanita pantherina*: under birch; mycorrhizal
Honey Fungus, *Armillaria ostoyae*: under birch; parasite
Cep or Penny Bun, *Boletus edulis*: under birch; mycorrhizal
Peppery Bolete, *Chalciporus piperatus*: under birch; mycorrhizal
The Miller, *Clitopilus prunulus*: under birch; saprophyte
Common Bird's Nest, *Crucibulum laeve*: on very rotten wood; wood decomposer
Hoof Fungus or Tinder Bracket, *Fomes fomentarius*: on birch; parasite
Wood Hedgehog, *Hydnum repandum*: under birch; very common, said to be edible
False Chanterelle, *Hygrophoropsis aurantiaca*: under birch; litter and wood decomposer
Brown Rollrim, *Paxillus involutus*: under birch; mycorrhizal, with variety of species
Razorstrop Fungus or Birch Polypore, *Piptoporus betulinus*: on birch; parasite

**Phycomycetes**
*Syzygites megacarpus*: on bolete, probably *Chalciporus*; parasite on boletes

**5. Tressady Wood, NC6904, visited 16.10.06. Mature oakwood, on east bank of Lettie River; grazed by sheep, ground vegetation sparse.**

**Basidiomycetes**
Red Cracking Bolete, *Boletus chrysenteron*: under oak; mycorrhizal
Small Stagshorn, *Calocera cornea*: dead wood, probably oak; wood decomposer
Purple Stocking Webcap, *Cortinarius stillatitius*: under oak; mycorrhizal
Heath Waxcap, *Hygrocybe laeta* var. *flava*: pathside; normally seen in grassland
False Chanterelle, *Hygrophoropsis aurantiaca*: under oak; litter and wood decomposer
Amethyst Deceiver, *Laccaria amethystina*: under oak; mycorrhizal
The Deceiver, *Laccaria laccata*: under oak; mycorrhizal
Curry Milkcap, *Lactarius camphoratus*: under oak; mycorrhizal
Wood Blewit, *Lepista nuda*: under oak; mycorrhizal
Heath Navel, *Lichenomphalia umbellifera*: in moss; one of few lichenised basidiomycetes
Common Bonnet, *Mycena galericulata*: on dead wood, probably oak; wood decomposer

Grooved Bonnet, *Mycena polygramma*: on dead wood, probably oak; wood decomposer
Brown Rollrim, *Paxillus involutus*: under oak; mycorrhizal with wide range of species
Brittlegill, *Russula fragilis*: under oak; mycorrhizal
Oilslick Brittlegill, *Russula ionochlora*: under oak; mycorrhizal
*Radulomyces molaris*: on wood, oak; wood decomposer
Earthfan, *Thelephora terrestris*: on soil/litter; litter decomposer
Candlesnuff Fungus, *Xylaria hypoxylon*: on wood, probably oak; wood decomposer

**Ascomycetes**
*Hyaloscypha hyalina*: on wood, oak; wood decomposer, very common
*Lachnum niveum*: on wood, oak; wood decomposer, usually on oak
*Mollisia cinerea*: on wood, oak; wood decomposer
*Poculum firmum*: on wood, oak; wood decomposer

**6. Garbh-allt, hazel and birch woodland on west side of burn, NC7204/7304, visited 26.9, 29.10**

**Basidiomycetes**
Butter Cap, *Collybia butyracea*: in litter under birch, 7204, 29.10
*Exidia recisa*: on willow wood, 7304, 29.10
*Exidia repanda*: on birch wood, 7204 and 7304, 29.10
Crimson Waxcap, *Hygrocybe punicea*: beside track, 7204, 29.10
Scurfy Fibrecap, *Inocybe petiginosa*: under birch, 7204, 29.10
Fiery Milkcap, *Lactarius pyrogalus*: under hazel, 7304, 29.10
Birch Milkcap, *Lactarius tabidus*: under birch, 7304, 29.10
Common Puffball, *Lycoperdon perlatum*: underbirch/hazel, 7304, 29.10
Pipe Club, *Macrotyphula fistulosa*: on birch wood, 7204, 29.10
*Melampsora populnea*: rust on aspen leaves, 7204, 29.10
Yellowleg Bonnet, *Mycena epipterygia*: rotten wood/litter, 7304, 29.10
Grooved Bonnet, *Mycena polygramma*: on hazel wood, 7304, 29.10
Lilac Bonnet, *Mycena pura*: in litter, 7304, 29.10
Brown Rollrim, *Paxillus involutus*: under birch, 7204, 29.10
Bay Polypore, *Polyporus durus*: on willow wood, 7304, 29.10
Pale Brittlestem, *Psathyrella candolleana*: in litter, 7304, 26.9
Charcoal Burner, *Russula cyanoxantha*: under birch/hazel, 7304, 26.9
Blackening Brittlegill, *Russula nigricans*: under hazel, 7304, 26.9
Shooting Star, *Sphaerobolus stellatus*: dead wood, probably hazel, 7304, 29.10
Bleeding Oak Crust, *Stereum gausapatum*: on hazel wood, 7304, 29.10

Sulphur Knight, *Tricholoma sulphureum*: in litter, 7304, 26.9
Plums and Custard, *Tricholomopsis rutilans*: in litter, 7304, 26.9

**Ascomycetes**
Purple Jellydisc, *Ascocoryne sarcoides*: on hazel wood, 7304, 29.10
*Bisporella citrina*: dead wood, 7304, 26.9
*Diatrypella favacea*: on birch bark, 7304, 29.10
*Mollisia cinerea*: dead wood, 7304, 29.10
*Propolis farinosa*: dead wood, 7304, 29.10
Candlesnuff Fungus, *Xylaria hypoxylon*: dead wood, 7304, 29.10

**Coelomycetes**
*Calcarisporium arbuscula*: parasite, on *Mycena* sp., 7304, 26.9

## 7. Other records of fungi made at a number of localities visited more briefly, in and around the village, and in the vicinity of Little Rogart Loch

**Basidiomycetes**
Honey Fungus, *Armillaria gallica*: roadside/woodland, Millnafua bridge, 7302, 29.10
*Dacrymyces minor*: on dead wood, probably hazel, south bank of Garbh-allt, 7202, 26.9
*Exidia recisa*: on willow wood, south bank of River Fleet, 7201, 26.9
Tinder Bracket or Hoof Fungus, *Fomes fomentarius*: on birch, woodland above Corry, 7302, 5.6 (coll. IME)
*Hemimycena crispata*: on dead grass, old sheep market, 7201, 5.6 (coll. IME)
Scarlet Waxcap, *Hygrocybe coccinea*: near Little Rogart Loch, 7203, 16.10
Parrot Waxcap, *Hygrocybe psittacina*: near Little Rogart Loch, 7203, 29.10
*Hyphodermella corrugata*: on birch wood, woodland above Corry, 7302, 5.6 (coll. IME)
Rufous Milkcap, *Lactarius rufus*: under birch, near Little Rogart Loch, 7203, 29.10
Brown Birch Bolete, *Leccinum scabrum*: under birch, near Little Rogart Loch, 7203, 16.10
Wood Blewit, *Lepista nuda*: in soil, rough grass, near Little Rogart Loch, 7203, 29.10
Stump Puffball, *Lycoperdon pyriforme*: spectacular group in mossy grass over large boulder, roadside near Millnafua Bridge, 7302, 29.10

Yellowleg Bonnet, *Mycena epipterygia*: in litter, south bank of Garbh-allt, 7202, 26.9
*Mycena meliigena*: on alder bark, south bank of River Fleet, 7201, 26.9
Willow Bracket, *Phellinus igniarius*: on willow wood, south bank of River Fleet, 7201, 26.9
*Phragmidium rubi-idaei*: rust, on raspberry, roadside, Millnafua Bridge, 7302, 29.10
*Puccinia annularis*: rust, on wood sage, south bank of Garbh-allt, 7202, 26.9; roadside quarry east of village, 7401, 16.10
Hairy Curtain Crust, *Stereum hirsutum*: dead wood, south bank of Garbh-allt, 7202, 26.9
Yellow Brain, *Tremella mesenterica*: on gorse, south bank of River Fleet, 7201, 26.9; obligate fungal parasite

**Ascomycetes**
Purple Jellydisc, *Ascocoryne sarcoides*: dead wood, south bank of Garbh-allt, 7202, 26.9
*Bisporella citrina*: on willow wood, south bank of River Fleet, 7201, 26.9
*Fimeria hepatica*: on cow dung, south bank of River Fleet, 7201, 26.9
*Hymenoscyphus scutula*: on dead nettle stem, south bank of River Fleet, 7203, 26.9; on dead herbaceous stem, near Little Rogart Loch, 7203, 16.10
*Lachnum niveum*: dead wood, south bank of River Fleet, 7201, 26.9
*Melanomma pulvispyrius*: dead wood, probably birch, near Little Rogart Loch, 7203, 16.10
*Microsphaera alphitoides*: powdery mildew, on oak leaves, south bank of Garbh-allt, 7202, 26.9
*Poculum firmum*: on oak wood, south bank of Garbh-allt, 7202, 26.9
*Rhytisma acerinum*: on sycamore leaves, by War Memorial, 7202, 29.10
*Taphrina populina*: on aspen leaves, south bank of Garbh-allt, 7202, 26.9
*Taphrina sadebackii*: on alder leaves, Rogart Church, 7104, 30.8.2004 (coll. IME)
*Taphrina tosquinetii*: on alder leaves, beside War Memorial, 7202, 29.10
Candlesnuff Fungus, *Xylaria hypoxylon*: on hazel wood, south bank of Garbh-allt, 7202, 26.9

**Phycomycetes**
*Pilobolus crystallinus* var. *kleinii*: on cow dung, south bank of River Fleet, 7201, 26.9

# Lichens

*Anthony Fletcher*

## Introduction

Lichens lie everywhere in the Highlands. In some places they are the dominant vegetation. Yet, they are strangely neglected by most naturalists. Why is this? It may be due to their reputation for being difficult to identify, needing a hammer and chisel to collect, examination under the microscope and use of strange chemicals to detect the species, and most have only Latin names. However, this neglect is unjustified, as many of the larger species can be collected just like flowering plants, they are quite easy to identify from a reliable book, and being mostly two-dimensional, are very easy to photograph.

Environmentalists increasingly value lichens because they are such excellent indicators of environmental conditions. It is possible, from one brief visit, to list the lichens and suggest how much air pollution the site is getting. One can also use them to detect climate changes, and the past history of a site. So, looking at lichens is a most useful habit to acquire.

Lichens are often considered to be the prime examples of mutualistic symbiosis. That is the living together of two dissimilar organisms for mutual benefit. In this case, the partners are a fungus with an alga, or in some cases, a cyanobacterium (formerly known as blue-green algae). Some 95% of the thallus ('plant') is fungus, with the rest algal cells. If a lichen is wetted and the surface scraped with a fingernail, the green layer of algae can be seen beneath the often highly coloured cortex.

Experiments have shown that the algae make foodstuffs such as simple sugars, which leak to the fungus, which absorbs them and uses them as food to grow. Close control is kept by the fungus, limiting the growth of the algae and stopping them from reproducing. In a lichen, it is only the fungus that can reproduce, and this is by ascospores which are made in characteristic tiny cups, saucers or cartwheels on the upper surface, which are often highly coloured. The ascospores, which can be seen only under a microscope, are released in wet conditions and are dispersed by water or on the legs of insects or birds. How an ascospore survives while it awaits an algal cell is unknown – the birth of a lichen is an unsolved problem.

Lichens come in many forms. We often notice the tufted, yellow-green beard lichens (*Usnea, Ramalina*) on tree branches. The colour comes from a chemical called usnic acid which is a known antibiotic and has recently been proposed as an anti-cancer drug. Most lichens produce chemical substances whose function is unknown, but which are useful for identification of species. Other types are leafy, but lie flat on the rock or tree bark, and are most often grey (*Parmelia, Physcia*), but can be yellow-green, brown or even black. The most difficult to identify are the crustose species, which lie like paint on a rock or bark surface. These have to be collected from rock with a hammer and chisel, or a sheath knife when on bark.

A number of lichens grow on soil and compete with flowering plants. The most conspicuous are the 'pixie cups' and 'reindeer lichens' (*Cladonia*), which are common on the moorlands around Rogart. An unusual group of species can be found on rocks in streams, where they are often completely submerged during spate (*Verrucaria*). Strangely, some can live on glass, wood or metalwork. The metalliferous lichens accumulate high quantities of metals and can appear rust-coloured (*Acarospora sinopica, Rhizocarpon oederi*).

These interesting organisms (we can't call them plants any more as Fungi are now considered to be a completely separate group from green plants and animals) have often been used by man, especially in the Highlands. Tree Lungwort (*Lobaria pulmonaria*), which looks like brown, dried-up lung, was once used as a medicine for lung diseases and coughs. The Dog Lichens on soil (*Peltigera*) were used to cure hydrophobia following a bite from a rabid dog. Whether these remedies worked is not known, but if they are the best you have, surely they are worth a try. Others, such as Cudbear (*Ochrolechia tartarea*) and Black Crottle (*Parmelia omphalodes*) were scraped off rocks and exported in barrels as a dye; a large industry in 19$^{th}$ century Glasgow. However, as many lichens are declining and threatened, it is not recommended to exploit them these days.

Returning to the present, lichens are good indicators of air pollution. The Rogart area is largely unspoiled by sulphur dioxide pollution from burning oil and coal, so the lichen flora is still quite rich. However, a modern form of air pollution is nitrogen, especially in the form of ammonia. This comes from fertilizers and intensive animal-rearing, around cattle, pig and chicken farms and in slurries. Lichens to look out for are *Xanthoria polycarpa*, *X. ucrainica* and *Physcia tenella*, which reflect this pollutant, and seem to be locally common in the Rogart village centre and around farms. However, many of the *Parmelia* species seen were fertile, reflecting the generally healthy conditions.

Away from air pollution, a distinct maritime influence can be seen on granite walls, where a number of seashore lichens are benefiting from seawater spray being carried inland from the coast. Woodland trees, such as Oak and Hazel, were fairly scarce in the parts of the parish looked at [although they do occur further afield], which limited the numbers of tree lichens found close to the village. But an idea of the lichen richness that may once have been more widespread is given by the Ash and Sycamores along the banks of the River Fleet. Luxuriant growths of *Ramalina fraxinea* [see illustration] there testify to the low sulphur dioxide content of the air. Juniper scrub is an unusual feature of the area, but the large bushes were strangely lacking in lichens. Perhaps they were too overgrown in former times and lichens failed to establish on them. Fires are also an important factor; there were several instances of rocks that had been completely stripped of their lichen cover.

The list that follows was made on a single visit, on $8^{th}$ May 2006, in the company of Ian Evans. Some 210 taxa were recorded, a few as yet unidentified. The area shows an interesting mixture of habitats. Some have lichens which prefer cool, moist, humid places, while others are warmer, dry and sheltered, especially below overhangs. The lichen flora generally has an interesting mixture of species which predominate in the Highlands, together with some more typical of southern England.

**Sites studied**

In one day, only a small part of the parish was visited, extending from Davochbeg, to the River Fleet at Davoch Bridge, through the village centre and then eastwards along the Corry Burn, with a visit to the juniper-rich areas and crags above Corry, all less than half a mile from the crossroads. The recording stations are grouped into local areas. For cross-referring, the sites are numbered and the numbers appear after the names of each species in the table.

1. Village centre, roof of public lavatory, NC72620196: slate, dominated by yellow-green *Rhizocarpon geographicum*.

2. Former Sheep Market, between road and railway, south of Corry Burn, NC72640195-72630189. Long-disused old sleepers, softwood posts and rails, spectacularly covered with dense lichens. Particularly noteworthy were *Imshaugia aleurites*, a specialist of pinewood, *Stereocaulon* and *Cladonia* spp., which normally grow on rock and soil respectively, and tree species such as *Pseudevernia furfuracea, Bryoria fuscescens, Parmelia* spp. and the beard lichen *Usnea hirta*.

3-8. South side of Corry Burn. The trees had lichens typical of wayside trees in sheltered and humid conditions. Stones in the river had aquatic *Verrucaria* lichens.

    3. Sycamores, on south-east bank of Corry Burn, NC72570192: sheltered and humid.
    3a. On rocks in river below 3: very shaded and sheltered.
    4. Alder, on south-east bank of Corry Burn, NC72570192: sheltered and shaded.
    5. Bird Cherry, on south-east side of Corry Burn, south-west of railway line, NC72540189.
    6. Sallows, on south-east side of Corry Burn, NC72500188.
    7. Ash, on south-east side of Corry Burn, NC72500188.
    8. Alder, at junction of Corry Burn and River Fleet, NC72500188.

9. Sycamore, on bank of River Fleet, NC72520184: well-lit, dominated by very large *Ramalina fraxinea* and other macrolichens.

10. Level crossing over railway, iron railings on north-east side, NC72530191. The iron railings (10a) were very exposed and sunny, with lichens on the uppermost surface. Some of these, such as *Acarospora sinopica* [see illustration] and *Rhizocarpon oederi*, were coloured rusty-red through accumulation of iron. Stones under the railings (10b) had lichens benefiting from water dripping from them.

    10a. Iron railings.
    10b. Stones under railings.

11-12. The walls of the watermill beside the Corry Burn and adjacent building, NC725019 (granite with lime mortar). The very sunny stones (11) had maritime species, picking up small amounts of seawater spray, such as *Caloplaca arnoldii*. Several bright orange species of *Caloplaca* and *Xanthoria elegans* were also found. The north-west-facing walls (12) also had abundant rock lichens.

11. South-east-facing walls of watermill, NC72540192.
12. North-west-facing walls of adjacent building (shady), NC72530194.

13-17. Roadsides between the Station and main road (A839). Sycamores and Ash bore characteristic lichens of well-lit bark and twigs. The picnic bench top, of pine, once treated with preservative but now weathered, had a good flora of 21 species, mostly of tree bark, though *R. geographicum* is more typical of rock. Granite wall tops had slow-growing crustose lichens receiving some bird lime.

13. Sycamore, on south-east side of road to level crossing, NC72570198 (gardened area).
14. Ash, by picnic bench, NC72580198.
15. Ash, by picnic bench, twigs, NC72590199.
16. Top of picnic bench, NC72590198.
17. Granite wall top (bird perch), garden, north-west side of road, NC72600202.

18-19, 22-24. Corry Burn, east of main road (A839). Some large Oak trees but rather shaded. The pin-head lichen *Calicium viride* specialises in dry crevices and together with the bright yellow *Chrysothrix candelaris*, spreads along only the coastal strip of East Sutherland. Even tiny twigs were colonised by minute black dots of *Mycoporum quercus*. Beard lichens (*Usnea* spp.) were abundant here on branches. The deeper ravine (sites 22-24) had Rowan and Hazel, with different lichens to those on Oak. These lichens, typical of base-rich bark, include *Peltigera*, *Nephroma* and fruits of *Graphis scripta* resembling Chinese writing.

18. Oak, south side of Corry Burn, NC72660199.
19. Oak, south side of Corry Burn, NC72670201.
22. Woodland, between road and Corry Burn, on hazel, NC72730204.
23. Ditto., on rowan, NC72750203.
24. Ditto., NC72770205.

20-21. Walls on roadside, north side of Corry Burn. These were of a pink granite, with quartz. The garage wall was mortared, encouraging lichens which prefer base-ions. This wall had a good lichen flora with some maritime elements such as *Caloplaca arnoldii*.

20. Mossy wall, north side of Corry Burn, NC72650201.
21. Wall of garden/garage on north side of road, NC72670202.

25, 27, 29. Open, south facing hillside above Corry, overlooking the village. The grassy pasture had low Junipers, spreading horizontally. Lichens on these were scarce, probably because the bark peels away and the lichens cannot maintain a foothold. This is an unusual habitat, with scarce lichens such as *Xylographa abietina*, which specialises in decorticate conifer wood. Several specimens could not be identified. A noteworthy species was the very rare *Caloplaca sorediella*, first described in December 2006; it is also known westwards on the Assynt coast, thence southwards to Wales.

25. Juniper, on open hillside, NC72800220.
27. Juniper, NC72910225 (specimens in with those from site 25).
29. Juniper, NC72820232.

26, 28, 30-36. Rock cliffs and outcrops above the Juniper area. These south-facing cliffs were densely overgrown with Gorse and heavily affected by fires. A good *Cladonia* flora was found on soil between rocks, including the heathland species *Cladonia uncialis*. Upland species were present, such as *Ophioparma ventosa* and *Parmelia discordans*. Crag 30 had several maritime lichens benefitting from seawinds.

26. Rock outcrop, NC72800220.
28. Rock outcrop, NC72890228.
30. Crag, NC72750237 (exposed; maritime influences).
31. Crags, NC72720238.
32. Boulder, NC72700241.
33. Boulders (not burned, unlike some nearby), NC72650243.
34. Crag, south-facing overhang, NC72660244.
35. Boulders, NC72650240.
36. Lichen heath on flat outcrop, NC72680236.

37. Large, moribund Elder in grassy pasture, NC72800226. Much of the bark was lost but a rich lichen flora persisted on the remaining bark, with species characteristic of Elder, such as *Caloplaca cerina*, *Lecania cyrtella* and *Lecanora sambuci*. *Calicium salicinum* is an eastern seaboard species. *Phaeophyscia endococcinea* is rather scarce.

38-40. Davochbeg, a large farm in the River Fleet floodplain. Lichens were found typical of nutrient-enriched conditions, from livestock urine and manure, which affected the lichens on walls and buildings. However, the wind-exposed position allowed many maritime species to exist, such as *Caloplaca maritima*. *Lauderlindsaya borreri* is a parasite on *Normandina pulchella*. Strangely, the host lichen was much less conspicuous than its parasite and in very poor condition. The sheltered garden on the north-east had Ash, Beech, Elms and Sycamores with quite different lichens to those in the nutrient-enriched south-west. Of interest was a pine telegraph pole bearing the crustose *Lecanora conizaeoides* which is very tolerant of atmospheric sulphur dioxide. It probably was introduced when the pine post was imported, perhaps from an air-polluted area.

38. Davochbeg, NC72310161.
38a. On granite walls of barn etc.
38b. On Elm in farmyard.
39. Davochbeg, walled garden, NC72310165.
39a. On Beech.
39b. On Sycamore.
39c. On Ash.
39d. On electricity pole.
40. Davochbeg, avenue of trees on north-east side of walled garden, NC72340165.
40a. On Elm.
40b. On Sycamore.

**The Lichen Species.** Species are named after the current Checklist, British Lichen Society website (www.theBLS.org.uk).

| Species | Sites | Notes |
| --- | --- | --- |
| Acarospora sinopica | 2 | A specialist of metal-rich rocks |
| Acarospora spp. | 2 | |
| Acarospora fuscata | 17 | |
| Anaptychia runcinata | 34 | A maritime species on rock |
| Anisomeridium polypori | 39 | On Ash |
| Arthonia arthonioides | 19 | Smooth bark |
| Arthonia radiata | 6, 7, 19, 37, 39a, 40b | Smooth bark |
| Arthopyrenia salicis | 3 | Smooth bark and twigs |
| Aspicilia caesiocinerea | 17, 21, 28, 30 | Exposed rocks |
| Athelia arachnoidea | 39c | Parasite on Xanthoria parietina |
| Bryoria spp. | 23 | Grey thallus cf. B. subcana |
| Bryoria fuscescens | 2, 16, 23, 27 | On bark, pales and rocks |
| Buellia griseovirens | 2, 16, 40b | On pales |

| | | |
|---|---|---|
| Buellia punctata | 2, 16 | On pales and nutrient enriched wood |
| Buellia stellulata | 21, 31 | On rocks |
| Calicium salicinum | 37 | On decorticate branch |
| Calicium viride | 18 | In crevices on Oak bark |
| Caloplaca spp. | 16 | On bench top, with K+ purple soralia |
| Caloplaca arnoldii | 11, 12, 21, 28, 30, 34, 38a | Usually on maritime rocks |
| Caloplaca cerina | 37 | Specialist of base-rich bark, esp. Elder |
| Caloplaca cerinelloides | 15, 37 | Elder twigs |
| Caloplaca citrina | 11, 37, 38a | On mortar |
| Caloplaca flavocitrina | 11, 38a | On mortar |
| Caloplaca flavovirescens | 2 | On concrete |
| Caloplaca maritima | 38a | A maritime rock species |
| Caloplaca saxicola | 11, 38a, 38b | On mortar, rare on Elm bark |
| Caloplaca sorediella | 25 | On rock, rare maritime spp. |
| Candelariella aurella | 16 | Nutrient-enriched sites |
| Candelariella reflexa | 14 | Probably a nitrogen pollution indicator, increasing in cities |
| Candelariella vitellina | 1, 2, 10a, 16, 21, 31, 38a | Common nitrogen pollution indicator |
| Catillaria chalybeia | 12, 16, 21 | On rock |
| Catillaria spp. | 6 | On Sallow |
| Cetraria chlorophylla | 2, 16 | On bark and wood |
| Chrysothrix candelaris | 18, 24 | On humid Oak bark |
| Cladonia chlorophaea | 20, 32 | Soil |
| Cladonia ciliata var. tenuis | 28 | Soil |
| Cladonia diversa | 28 | Acid soil |
| Cladonia fimbriata | 18, 19, 25, 29, 37 | Soil and wood |
| Cladonia furcata | 28, 35 | Acid soil |
| Cladonia macilenta | 29 | Acid soil |
| Cladonia portentosa | 2, 28 | A reindeer lichen of acid soil and peat |
| Cladonia pyxidata | 28 | On soil |
| Cladonia rangiformis | 26, 28 | On soil |
| Cladonia subcervicornis | 25, 28, 31 | On soil |
| Cladonia subulata | 2 | On dusty soil |
| Cladonia uncialis | 36 | On wet peaty soil |
| Cliostomum griffithii | 6, 8, 25, 27, 39, 40a | On smooth bark and Juniper |
| Cornicularia aculeata | 28 | On soil in rock crevices with Cladonia |

| | | |
|---|---|---|
| Dermatocarpon miniatum | 30 | On seepage rocks with maritime influence |
| Diplotomma alboatrum | 11, 40a | On base-rich rocks, mortar and Elm bark |
| Dirina masssiliensis f. sorediata | 25 | On shaded rock |
| Evernia prunastri | 2, 3, 18, 19, 25, 37, 39a | On bark and twigs |
| Fuscidea cyathoides | 2, 21, 28, 31 | On siliceous rocks |
| Graphis scripta | 22 | On shaded, smooth bark |
| Haematomma elatinum | 2 | On shaded palings |
| Haematomma ochrolecum var. ochroleucum | 32, 38a, 40a | Shaded rock overhangs |
| Haematomma ochroleucum var. porphyrium | 34 | Shaded rock overhangs |
| Hypogymnia physodes | 2, 16, 18, 19, 23, 25 | On pales, rock and bark, fertile at 23 |
| Hypogymnia tubulosa | 2, 25, 28 | On pales, rock and bark |
| Imshaugia aleurites | 2, 8 | On pales and acid bark |
| Indet. | 25 | On soil, photobiont Gloeocapsa (Cyanobacteria) |
| Lauderlindsaya borreri | 39 | On Ash, parasite on Normandina pulchella |
| Lecania cyrtella. | 37, 39 | On Elder & Ash |
| Lecanora spp. | 37 | On Elder |
| Lecanora cf aitema | 3 | On sheltered Sycamore |
| Lecanora albescens | 11, 21, 38a | On mortar |
| Lecanora campestris | 11 | On siliceous wall stones |
| Lecanora carpinea | 3, 4, 6, 37 | On Elder bark and twigs |
| Lecanora chlarotera | 3, 4, 6, 19, 25, 27, 37 | On bark |
| Lecanora confusa | 2, 4 | On pales |
| Lecanora conizaeoides | 39d | Sulphur dioxide pollution indicator on wood and bark |
| Lecanora dispersa | 11, 38a | On mortar |
| Lecanora expallens | 4, 13, 18, 19, 25, 39a | On tree bark and twigs |
| Lecanora gangaleoides | 10b, 12, 21 | On rocks |
| Lecanora intricata | 2, 10a, 21 | On rocks and wall tops |
| Lecanora orosthea | 31 | Shaded overhangs |
| Lecanora persimilis | 3, 15, 37, 39c | Small twigs |
| Lecanora polytropa | 10a, 12, 32 | Rocks |
| Lecanora praepostera | 21 | Shaded rock crevice |
| Lecanora saligna | 2, 16, 38a, 39d | Pales and wood |

| | | |
|---|---|---|
| Lecanora sambuci | 37 | On Elder |
| Lecanora sulphurea | 30 | Sunny rocks |
| Lecanora symmicta | 2, 4, 7, 25, 27 | On pales and twigs & Juniper |
| Lecanora varia | 2 | On pales |
| Lecidea cf. nylanderi | 25 | On Juniper bark |
| Lecidea lactea | 10b, 21, 31 | On rocks |
| Lecidella elaeochroma | 3, 6, 19, 37, 40b | Wood, bark and twigs |
| Lecidella elaeochroma f. soralifera | 25 | On Juniper |
| Lecidella scabra | 21, 28 | On rocks |
| Lecidella sp. | 25 | On Juniper |
| Lecidella stigmatea | 12, 21 | On mortar |
| Lepraria incana | 17, 19, 39, 40a | Crevices in bark |
| Lepraria neglecta agg. | 31 | Open, peaty rocks |
| Leprocaulon microscopicum | 30 | In rock crevices, often maritime |
| Leproloma lobificans | 18, 31 | On soil |
| Leptogium gelatinosum | 2 | On concrete of loading bay |
| Leptogium teretiusculum | 34 | In mossy seepage track |
| Leptorhapis maggiana | 25 | On twigs of Juniper |
| Lichenoconium lecanorae | 4, 6, 7, 37 | Parasite on Lecanora carpinea, L. symmicta, L. chlarotera |
| Micarea denigrata | 2 | On pales |
| Micarea nitschkeana | 37 | On small Elder twigs |
| Mycoporum quercus | 19 | On small Oak twigs |
| Nephroma laevigatum | 22 | On mossy Hazel bark |
| Normandina pulchella | 39 | On Ash, heavily parasitised by Lauderlindsaya |
| Ochrolechia spp. | 25 | White sorediate crust, dense on Juniper bark |
| Ochrolechia androgyna | 2 | On pales |
| Ochrolechia parella | 20, 28, 30, 39c | On rocks and walls |
| Ochrolechia subviridis | 6, 27, 38b | On various tree bark |
| Ochrolechia tartarea | 35 | On boulder in grassland |
| Ochrolechia turneri | 2 | On pales |
| Opegrapha atra | 23 | On Rowan |
| Opegrapha herbarum | 23 | On Rowan |
| Opegrapha varia | 39 | On Ash |
| Opegrapha vulgata agg. | 24 | On Hazel |
| Ophioparma ventosa | 33 grey form | On boulder, grey form lacking usnic acid |
| Ophioparma ventosa | 35 yellow form | On boulder, yellow form with usnic acid |

| | | |
|---|---|---|
| Parmelia conspersa | 28 | On rocks |
| Parmelia delisei | 30 | On rocks |
| Parmelia discordans | 32 | On boulder |
| Parmelia glabratula ssp. fuliginosa | 2, 21, 27, 30 | On rocks |
| Parmelia omphalodes | 28, 31, 32 | On rocks |
| Parmelia saxatilis | 2, 6, 10b, 21, 25, 28, 30, 39a | On rocks, walls, pales |
| Parmelia subaurifera | 3, 4, 6, 15, 16, 19, 25, 37, 39a | On bark and twigs & Juniper |
| Parmelia subrudecta | 5 | On sheltered, smooth bark of Cherry |
| Parmelia sulcata | 2, 3, 6, 16, 18, 23, 25, 37, 39a | Common on bark, twigs and pales |
| Parmelia ulophylla | 14 | On sheltered Ash branch |
| Peltigera collina | 23 | On mossy bark |
| Peltigera didactyla | 25 | On soil among Juniper |
| Peltigera lactucifolia | 28 | On soil among rocks |
| Peltigera membranacea | 26, 28 | On soil among rocks |
| Peltigera praetextata | 22 | On mossy Hazel bark |
| Peltigera rufescens | 26 | On soil among rocks |
| Pertusaria albescens | 39a | On Beech bark |
| Pertusaria corallina | 28, 31 | On rocks, 31 with parasite |
| Pertusaria excludens | 33 | On boulders among grassland |
| Pertusaria hemisphaerica | 6 | On rough Sallow bark |
| Pertusaria pertusa | 19, 39a, 40b | On young bark of various trees |
| Pertusaria pseudocorallina | 28, 33 | On rock |
| Phaeophyscia endococcinea | 37 | A scarce species on old Elder |
| Phaeophyscia orbicularis | 15, 37, 38b | On base-rich bark |
| Phlyctis argena | 6, 39a, 40a | Shaded tree bark |
| Physcia adscendens | 37, 39c | Nitrogen loving species on tree bark, twigs and rocks |
| Physcia aipolia | 6, 7, 15, 37, 39a, 39c | On Ash and Elder twigs |
| Physcia adscendens | 37 | On Elder |
| Physcia caesia | 1 | On sunny slate roof |
| Physcia tenella | 6, 7, 15, 37, 39c | Nitrogen loving species on tree bark, twigs and rocks |
| Physcia tribacea | 40a | Scarce species on Elm bark |
| Physconia distorta | 6, 7, 9, 14, 16, 37, 39a, 39b | Common on mossy bark |

| | | |
|---|---|---|
| Physconia grisea | 13, 14, 37, 39a | On rock and bark |
| Placynthiella dasaea | 31 | On acid soil |
| Placynthiella icmalea | 16 | On wood and pales |
| Placynthium nigrum | 21 | On mortar |
| Platismatia glauca | 2, 27, 35 | On twigs and sunny boulders |
| Porina chlorotica | 20, 21 | In water seepage rocks |
| Porina lectissima | 34 | In seepage track |
| Porpidia macrocarpa | 25, 31 | On shaded rock |
| Porpidia tuberculosa | 10a, 21 | On rock and ironwork |
| Protoblastenia rupestris | 11 | On mortar |
| Pseudevernia furfuracea var. ceratea | 2 | On sunny pales; containing olivetoric acid – C+ red |
| Ramalina calicaris | 9, 40b | On tree bark and branches |
| Ramalina farinacea | 4, 37, 39a | On Alder |
| Ramalina fastigiata | 4, 9, 15, 37, 39a | On tree bark and branches |
| Ramalina fraxinea | 5, 9, 37, 39a | On sunny tree trunks |
| Ramalina subfarinacea | 35 | On sunny rock |
| Rhizocarpon cf. lecanorinum | 31 | On cliff rocks |
| Rhizocarpon concentricum | 21 | On mortar |
| Rhizocarpon geographicum | 1, 10a, 16, 21, 28, 30, 31 | On rocks |
| Rhizocarpon hochstetteri | 12, 17, 21 | On rocks |
| Rhizocarpon cf. intermediellum | 25 | On rocks |
| Rhizocarpon obscuratum | 1, 12, 16, 21 | On rocks |
| Rhizocarpon oederi | 10a | On ironwork |
| Rhizocarpon richardii | 30 | Maritime species on rocks |
| Rinodina sophodes | 3 | On Ash twigs |
| Sarcogyne regularis | 11 | On mortar |
| Schaereria fuscocinerea | 31 | On rocks |
| Scoliciosporum chlorococcum | 39c | Pollution tolerant, on farmyard Ash |
| Scoliciosporum umbrinum | 2, 10a | On pales and rocks |
| Sphaerophorus globosus | 28 | On soil between rocks |
| Stereocaulon leucophaeopsis | 2 | On pales |
| Stereocaulon vesuvianum | 2 | On pales |

| | | |
|---|---|---|
| Sticta limbata | 22 | On mossy Hazel |
| Tephromela atra | 2, 11, 21, 28, 30, 38a | On rocks |
| Trapeliopsis flexuosa | 2 | On pales |
| Trapeliopsis granulosa | 2, 25, 29, 37 | On wood and pales |
| Trapelia involuta | 25 | On rock |
| Tremolecia atrata | 10a, 21 | On rocks, usually metal-rich |
| Usnea fulvoreagens | 23 | On Rowan |
| Usnea hirta | 2, 25 | On twigs |
| Usnea cf pendulina | 23 | On Rowan bark |
| Usnea spp. | 16 | Too small to identify, on picnic bench |
| Usnea subfloridana | 4 | On Alder |
| Usnea wasmuthii | 19, 23 | On Oak and Rowan branches |
| Verrucaria aquatica | 3a | A freshwater aquatic |
| Verrucaria macrostoma | 11 | On mortar |
| Verrucaria maura | 30 | On rocks, in seawater enriched seepage |
| Verrucaria muralis | 11, 38a | On mortar |
| Verrucaria nigrescens | 2, 12, 21 | On mortar |
| Verrucaria praetermissa | 3a | A freshwater aquatic |
| Xanthoria elegans | 11 | Scarce, sunny rocks |
| Xanthoria parietina | 3, 14, 15, 16, 21, 37, 38a, 38b, 39 | Common on base-rich tree bark, twigs and mortar |
| Xanthoricola physciae | 37 | Parasite on Xanthoria parietina apothecia |
| Xanthoria polycarpa | 3, 16, 25, 39c | On twigs, especially nitrogen enriched |
| Xanthoria ucrainica | 2, 4, 16, 38b | On tree bark |
| Xanthoria ulophyllodes | 2, 38b | On tree bark |
| Xylographa abietina | 25 | On Juniper bark |
| | | |
| Indet. sorediate crusts | 27 | Juniper bark |
| 'Verdigris' (on moss) | 39c | On mossy rock of garden wall |
| Trentepohlia aurea (algae) | 40b | On shaded Sycamore |
| Fungus indet. | 25 | On Juniper bark |
| Hysterium spp. (Fungi) | 25 | On Juniper bark |